ARIZONA
HIGHWAYS

Presents
Desert Wildflowers

DAVID MUENCH

Text by
Desert Botanical Garden Staff

Technical Editors
Gary Paul Nabhan
Assistant Director

Jane Cole
Librarian

Graphs, Illustrations, and Maps by
James R. Metcalf

Photography by
Arizona Highways Contributors

Contents

DON B. STEVENSON

JAMES TALLON

DON B. STEVENSON

Arizona Highways Presents Desert Wildflowers

Hugh Harelson — Publisher / **Wesley Holden** — Managing Editor / **James E. Cook** — Associate Editor / **James R. Metcalf** — Design and Production

Prepared by the Related Products Section of *Arizona Highways* magazine, a monthly publication of the Arizona Department of Transportation.

Library of Congress Catalog Number 87-51119
ISBN 0-916179-15-X

Printed in Japan

Preface

To those of us who have the privilege of working at the Desert Botanical Garden, wildflowers have become a part of our daily lives. Throughout the year, as each season succeeds the other, we watch diverse species put on their natural delights. And along with the aesthetic enrichment that these flowers give us comes deeper awareness and understanding of these desert plants—and the urgent need to protect them and their habitat.

From the thousands of wildflowers that grow on the desert, the 63 species described in this book represent a wide variety to include all four growing seasons. These plants are native to Arizona and frequently seen. For the reader's convenience, Desert Botanical Garden will provide an annual update list of places where they are available through commercial seed and plant suppliers. Taken as a group, they will get you started in knowing and growing wildflowers in all four seasons.

Gary Paul Nabhan's introduction will help you learn how to develop the special senses that are needed. Read on to find out what makes desert wildflowers special and why the flowering seasons in Arizona are so unpredictable.

This book will introduce you to the wildflower seasons: late winter/early spring; mid-spring; late spring/summer; and late summer/fall. Each flower is described and located on the regional map so that you know where and when to look for it. Graphs at the beginning of the season chapters show actual bloom times recorded at the Desert Botanical Garden.

But that's only the beginning. Chapters 8 and 9 tell you how to grow your own wildflower garden, harvest the flowers and fruit, and prepare your harvest for use. The planting guide, references, index and additional information at the end of the book fill out the story. In return, we would like to hear from you. Let us know how well your watching and growing goes, and what works well and what doesn't. We are happy to have your interest join ours in preserving and restoring this beautiful desert land.

We hope this book becomes a guide and companion in your quest for beauty and understanding.

Robert G. Breunig
Executive Director
Desert Botanical Garden

The short rainy seasons near Tucson in the Saguaro National Monument bring dramatic thunderstorms to the Sonoran Desert. WILLARD CLAY

Introduction
The call of the wildflower:
Heard in the wilderness and echoed in the backyard

W e listened hard, for we were being called out to Arizona's desert wilderness. We could hear it as the background music beneath the coyote's howl, the burrowing owl's call, and the singing of desert springs.

The sounds came from a distant choir gathered in the Pinta Sand Dunes in the Cabeza Prieta Wildlife Refuge one April. The spring winds rattled and rustled the blazingstars, snowy sunflowers, desert-marigolds, and lupines. The ribbony leaves of the ajo-lilies sang in the sand as they scoured out an arc around their trumpet-like bloom.

The desert blossoms were beckoning. We went out to meet the music. And yet, as we left our homes in Phoenix, Tucson, and Flagstaff, we had no way of knowing whether the wildflowers would still be showing by the time we reached them. Some make only a two- to three-week appearance in bloom before they wither. Other summer bloomers have only a six- to eight-week lifespan as annual plants before they die back. They then leave their seeds in the desert sand or their bulbs dormant below, mute until a good rain comes some other year.

Wildflowers, like the stones that naturalist Annie Dillard has written about, may or may not talk, let alone sing. And yet, we still gain something by tuning into them, by listening for their wildness.

When desert wildflowers do call, sometimes it is the visible equivalent of a high-decibel shout. At Picacho Peak on one occasion, the goldpoppy show was spectacular enough to keep highway traffic past it slowed to 10 miles an hour. At Kitt Peak you can watch the desert floor below break open into a kaleidoscope of colorful poppies, owl-clovers, verbenas, chias, and globemallows. On the Black Canyon Freeway rising towards New River from Phoenix, the brittlebush and paloverde can paint the volcanic hills a gaudy golden for weeks on end.

Wildflower hunters in Arizona have developed into an opportunistic breed, ready to roam at the hint of a floral show. I know several plant fanatics who will drive a hundred miles out of their way for a chance to lie down in the midst of a mixed stand of a dozen or so kinds of bellyflowers—the ones that must be less than a nose-length away for you to be able to identify their microscopic blossoms. Other friends suddenly arrange evening events when their night-blooming cereus decides to open.

Some of these spectacles move around, like mirages. Although most outsiders think of plants as organisms that "sit still," wildflower hunters claim that the best blooms move around as much as deer or antelope do.

But you don't have to drive a hundred miles, or even ten, to savor the beauty of Arizona's native plants. This was brought home to us by an elderly Indian friend of ours, a man who has never owned a car or traveled very far from his Sonoran Desert place of birth.

Every year, our families would visit at Easter time. With cane in hand, he would walk us around his yard, showing us the flowers that had volunteered, the others that he had planted, and still others tended by his wife.

He clearly showed favoritism, though. His imagination had been captured years ago by the beardtongue that he called "wind's flower" in his native tongue. The wind, he claimed, blew the seeds into his yard years ago. When he saw the plants rise, he watered them and weeded away competitors. The next year, when winter came, they arose again, but from an entirely different spot within his yard. Again he cared for them with the same fervor as if they were prize roses he'd bought from a nursery. The following year, he began to walk his yard earlier in the season, curious to find out where they would pop up next.

Soon, he began selecting variants, encouraging whitish mutants and multi-lobed beardtongues to let their seeds be shed more abundantly. By the time our friend had gone blind with age, his yard had become a "menagerie" of the wind's flowers.

And the memory of their beauty has not disappeared from his mind's eye.

The beauty of native wildflowers is a kind that you *can* take with you, but not merely by picking a bundle, or by snapshots; you can let their colors, shapes, and fragrances seep deeply into your life. This book is in praise of the cactus flower, the gourd blossom, the rust-colored seedheads of native grasses, and the tiny blooms of bellyflowers that make us stop and take notice of the earth around us.

It also is in praise of the people who have let these other organisms into their daily lives. And it is an aid to encourage you to do the same.

Gary Paul Nabhan
Assistant Director
Desert Botanical Garden

Chapter 1
Arizona wildflowers come in all shapes, sizes, and seasons

Something is in bloom in Arizona year-round. While the vegetation of many states is sitting under snow, our "spring" wildflowers are breaking into bud, at least as early as February. A succession of herbs, trees, and cacti flower up until the monsoon rains begin in late June or early July, then some of them flower again. Others add their color through the late summer and fall. If a hurricane fringe storm hits Arizona in October or early November, the blossoming and fruiting of herbs will be prolonged.

Even in moderately dry years, Decembers may come with brittlebush, fairyduster, and desert broom still in bloom. By the time the natural vegetation's harvest ebbs, gardeners should have already seeded their favorite native flowers in beds around their yards, supplementing them with water until the first winter rains begin.

This colorful yearly cycle offers no rest for the dedicated.

Perhaps blooming seems continuous because our definition of "wildflower" is so generous. In addition to the characteristic low-growing herbs with blossoms as lovely as any cut flower in a florist's shop, our wildflower list includes certain showy native trees and shrubs, cacti, and succulents.

Yet even the term "native" can be confusing. You can order so-called native plants through the mail that you may not be able to find in an undisturbed desert area near your home. They may still be native in the sense that the particular species has grown naturally in the Sonoran, Chihuahuan, or Mohave Desert areas of Arizona since prehistoric times. Nevertheless, the particular seed you order may have been propagated in a section of the Mohave Desert in California.

How are these "natives" any different from African daisies? To begin with, they are already adapted to Arizona weather and soil. They also may have reciprocally evolved with the soil microbes, pollinators, and seed carriers found in Arizona environments, so that all these elements favor, or benefit from, their presence.

Native plants give Arizona part of its identity, a distinctive signature by which visitors from other parts of the world can tell where they are. Arizona's natural environment is loved and respected by people from around the globe. We should maintain this integrity, rather than dilute it with exotic plants that add little to our sense of place.

If we can buy seed or nursery stock of these natives, to what extent are they truly wild? Simply put, wildflowers are any plants with showy blooms that have not been genetically bred by human kind. They can be the low-lying trumpet flowers of ajo-lilies, the millions of sweet pale clusters hanging from a mesquite tree, or the crown of cream blossoms atop a giant saguaro.

In their natural environment, wildflowers are capable of reproduction and persistence without man's intervention on their behalf. However, many respond well to any extra water or nutrients offered to them. The true test of their wildness is when the backyard-harvested seeds are broadcast back into a natural, untended environment. If they can be found volunteering in that place for many years to come, then they have not become as domesticated as our backyard vegetables and house plants have. In contrast, exotic ornamentals with the "threads" of their evolutionary history bred out of them will fail under the unpredictable conditions that Nature has to offer.

A wildflower population still has enough natural resilience to respond to the range of conditions characteristic of its homeland. It has not had its genetic variability narrowed by specialized horticultural breeding. Its seeds may not germinate all at once, for nature keeps wildflowers from "putting all their eggs in one basket." By having seedling emergence spread out over several weeks each year, and several years overall, a wildflower population can avoid extinction caused by one catastrophic event such as an unseasonable freeze or locust epidemic.

For those of us who find variety to be the spice of life, wildflowers satisfy our aesthetics in another way. Within the same species or stand, not all individuals are identical in size, shape, or color. Unlike plant breeders who develop hybrid agricultural crops, Nature does not seek uniformity. Scientists have documented that desert wildflowers are genetically diverse in terms of inherited traits such as flower color, seeds per fruit, or tolerance of salty soils. Near Casa Grande in the spring, travelers often notice white lupines sprinkled in among the true blue

(LEFT) *A native plant grouping of desert-marigold, strawberry hedgehog, and Engelmann's prickly-pear is on display near Superior at the Boyce Thompson Southwestern Arboretum.* JOSEF MUENCH

ones. This may be what ecologists mean when they say "Nature abhors monotony."

Combine this genetic variation with the vagaries of the desert environment, and wildflower lovers experience what scientists call "phenotypic plasticity." Wild plants are "plastic" or adaptable in the sense that they can flower in a drought year when they are the height of your handspan, but "stretch" to reach your waist when they flower in a wet year. Their size, shape, and maturity time are all *malleable* so that they can become fine-tuned to the impending conditions.

It is the interaction between enigmatic arid environments and indefatigable flowering plants that makes desert wildflower watching such a joyous gamble. No two years, no two places are exactly alike in their shows. In the toughest periods, the plants are stripped down like hot rods, to the bare essentials needed in their race with time. In more benign years, the desert burgeons with such a spread of color that you wonder how anyone could ever call it a barren wasteland.

Perhaps more than in any other place in North America, wildflowers in the desert are interesting not so much in and of themselves as for what they tell us about their environment. Here, where rainfall can vary fourfold from best to worst years in the same decade, annual wildflower heights can vary sixteenfold. Inundated by runoff, an island within a desert arroyo may have twenty times the plant growth of adjacent rocky slopes. And a low-lying basin receiving a heavier frost may have its flowering season shrunken while a butte nearby escapes the freeze to dazzle all beholders.

Keep in mind that there is not just *one* Arizona desert, and that each distinctive arid land vegetation type in the state keeps its own suite of wildflowers. The dominant vegetation of any place is shaped by extreme high and low temperatures, frequency of freezes or fires, seasonality of rainfall, and duration of drought. These factors also affect which annuals and small cacti grow beneath the tree and shrub canopies.

Much of the area we cover in this book falls within two Sonoran Desert types: the Arizona Uplands to the east and the Lower Colorado to the west.

The Arizona Uplands type is the best-known desert vegetation in Arizona. It is the one emphasized in this book, for its saguaro and paloverde association harbors many spring and summer annual wildflowers.

The Lower Colorado type supports many of the wildflowers peculiar to sand dune vegetation and has fewer tall cacti, trees, and succulents than the Arizona Uplands. North of the Lower Colorado, Mohave Desert

Wild grasses, fiddlenecks, owl-clovers, and jumping chollas light up the desert. DAVID MUENCH

Vegetation types

- Arizona Uplands type
- Lower Colorado type
- Mohave Desert type
- Chihuahuan type
- Chapparal type
- Semidesert Grasslands type
- Evergreen Woodland type
- Sky Islands

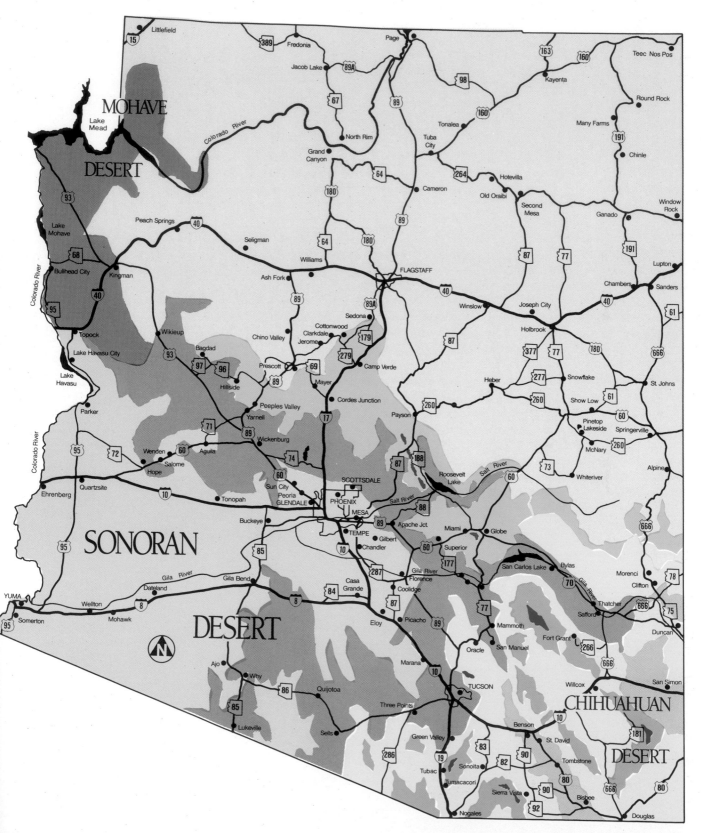

Adapted from D.E. Brown/C. H. Lowe, 1980.

Some of the most spectacular desert vegetation in the world grows in the Sonoran Desert in Arizona. But there is not just one Arizona desert. Each distinctive arid land vegetation type in the state keeps its own suite of wildflowers.

vegetation occurs in Arizona in the Kingman area. The Joshua-tree is the most well known plant of this vegetation, which is rich in spring wildflowers, but almost lacking in showy summer annuals. In southeastern Arizona, the fringe of the Chihuahuan Desert features many summer annuals, but lacks columnar cactus as well as tall trees.

In this book we also cover many wildflowers which extend up into semidesert grassland, plains grassland, and evergreen woodland at higher elevations in south-central and southeastern Arizona. We also cover some spectacular species of the chapparal and evergreen woodlands along the Mogollon Rim. Although most wildflowers range beyond more than just one vegetation type, the majority reach their optimal conditions for growth in a particular climate and vegetation type. If you are looking for them, knowing a little about the plants most commonly associated with them will cue you into their occurrence.

Different sets of wildflowers "read" these deserts in distinctive ways. One group, known as the "drought escapers," grow only during the time when conditions are least desert-like. While the ground is moist, they will grow so rapidly that they can mature in six to twelve weeks. Before drought has a chance to set in again, these wildflowers have bloomed, fruited and gone dormant or died; they have "escaped" death due to thirst.

Other plants persist through drier times of the year by drawing upon reserves of water not available to the annual wildflower. The best example of these plants, which tap hidden reservoirs of water, is the mesquite tree. Its roots along floodplains reach into the underground flow of soil moisture. Along permanent streams, mesquites never truly experience water stress. Also found in uplands or away from rivers, mature mesquite roots are often more than 30 feet deep, and have been found as far down as 150 feet! They are drawing upon a large root area for absorbing water and nutrients. It is not so surprising then that the plant productivity in a mesquite forest along a desert flood plain is the highest of any arid area in the world.

Cacti, agaves, and other succulents, such as tuberous wild gourd plants, draw upon water stored in their own tissue, rather than that deep in the ground. During dry periods, they close up shop, slough off their tiny root hairs, and minimize water loss. Within twenty-four hours of a drenching rain, their roots are reactivated, and root hairs are regrown in the upper layers of the soil. They

Text continued on page 25

Beavertail prickly-pear persists through drier times of the year by drawing upon reserves of water stored in its own tissue. Late spring to midsummer finds many blooms in the Kofa Mountains near Yuma.
DAVID MUENCH

(ABOVE) *Visitors to Organ Pipe Cactus National Monument enjoy the brightly colored penstemon, in association with numerous other plants, alongside the scenic loop drives.* WILLARD CLAY

(RIGHT) *Parry's agave is often called "century plant," because it takes many years to develop its flowering stalk. Here it is shown with a stray member of the sunflower family nestled in its center.* DAVID MUENCH

(FAR LEFT) *Rocks and sloping terrain often form a natural habitat for desert wildflowers. This barrel cactus is a natural associate of the cryptantha, cholla, and penstemon in their rocky home.* WILLARD CLAY

(LEFT) *Desert trees, such as the ironwood, often act as a nursery for tender, young saguaros which, many years later, outgrow or even kill their protective cover.* ROBERT CAMPBELL

(BELOW) *An evening-primrose, sprawled in a rocky spot, catches enough moisture for its brief bloom time.* LARRY ULRICH

(FOLLOWING PANEL, PAGES 16-17) *The century plant makes a spectacular showing when it blooms, adding beauty to the landscape of the Canelo Hills in the Coronado National Forest.* WILLARD CLAY

(ABOVE) *The Huachuca Mountains, background, are one of several sky islands that rise out of the desert in southern and southeastern Arizona.* WILLARD CLAY

(LEFT) *Wild morning-glory shows off its blue flowers in the Arizona deserts but is also native to elevations above 3,000 feet such as this setting in the Huachuca Mountains.* WILLARD CLAY

(BELOW) *Near Safford, an early spring field of gold-poppies gives a spread of color on the hillside.* DAVID MUENCH

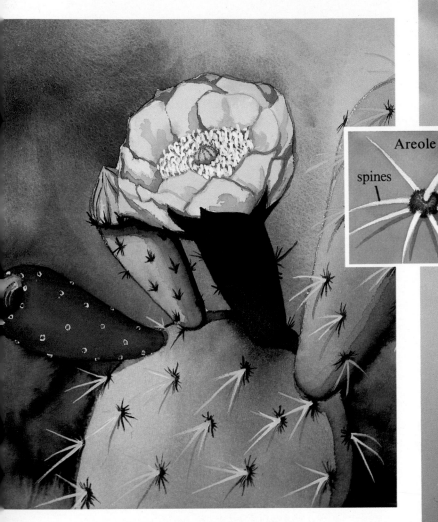

spines

Areole

glochids

(ABOVE) *A cactus has five characteristics.*

First, all cacti have areoles from which grow spines and often, minuscule hairs or glochids. Out of the areole develop two sets of buds, one for flowers, the other for spines and glochids. The absorbent button-like areole helps keep moisture in and heat out.

Second, all cacti are perennial, requiring more than one season to mature and flower.

Third, cacti usually have a wheel-shaped or funnel-shaped flower, with many sepals and petals. Below the narrow part of the floral tube, the ovary develops into a fruit.

Fourth, all cacti fruit have one compartment with seeds distributed throughout the pulp, like watermelon.

Fifth, all cacti seeds produce two embryo leaves following germination. ILLUSTRATIONS BY JAMES R. METCALF

(RIGHT) *The name "Joshua-tree" came from early Mormon settlers, since this yucca seemed to be spreading its arms in welcome to weary travelers. At the edge of the Mohave Desert, it blooms in abundant natural stands and is a popular cultivated plant in desert gardens and as highway or parkway plantings.*
JOSEF MUENCH

(LEFT) *The elegant and graceful ocotillo blooms in the Mohave Desert near Kingman, where it shares a habitat with creosotebush, cholla, yucca, and other wildflowers.* DAVID MUENCH

(BELOW) *Near Wickenburg, a group of riders enjoys the yucca forest along the trail. Individual stands of yucca bloom throughout the spring season and into summer.* JAMES TALLON

(ABOVE) *Prickly-pear blooms attract pollinators, photographers, and plant scientists. The variation of color within a single prickly-pear species confuses amateurs and fascinates propagators.* J. PETER MORTIMER

Text continued from page 11

quickly absorb this superficial soil moisture, their tissues swelling up with water supplies to tide them over the subsequent months. A barrel cactus can continue growth without any replenishment of its water stores for a month and a half after the soil becomes drier than the plant.

Some trees and shrubs drop their leaves during drought to minimize water loss, but still carry on some food-making activities via their photosynthetic bark. These drought tolerators include paloverde and ocotillo. Other trees and shrubs do not have massive stores of water to fall back on, and they are the true drought resisters—the ones which must bear the brunt of harsh conditions. They have only tiny leaves to begin with, often covered with resins or tiny gray hairs to reduce heat loading and water loss. Drought resisters like the creosotebush drop their leaves only under the most extreme water deficits but respond rapidly once rain comes again.

In addition to the survival strategy which each desert plant uses, other factors influence its flower size, color, scent, and timing. The availability of animal pollinators is one of the the strongest influences in shaping floral architecture, chemistry, and lifespan.

Each pollinator has special nutritional needs which are met by the kinds of sugars and other nutrients in the flowers it visits. For example, hummingbirds tend to favor long, tubular flowers that are red, purple, orange, or pink in hue, and have *sucrose* rich nectars. Bats specialize in pollinating night-blooming, pale, broadly tubular or conical flowers that are laden with nectars rich in *glucose* and *fructose*. Hawkmoths also prefer pale flowers that stand out at dusk, but favor strongly-scented flowers with *sucrose* dominated nectars concealed in deep narrow tubes. Butterflies also suck nectar from deep narrow corollas but may do so during the day and commonly choose flowers that are pink or other highly saturated colors.

There are some bees which specialize in a particular shape and size of flower but rarely favor pure red blooms. Many of the favorite bee-pollinated flowers have brightly-colored "landing pads" or "honey guides" which point the way to hidden nectar in the flower. Larger bumblebees and carpenter bees can open "closed" flowers or force themselves beneath masses of stamens to reach pools of nectar in the bases of flowers. In doing so, then escaping, they often end up covered with the pollen stuck on them after passing through the stamens and carrying this to other flowers. Moths and beetles also have special features which match them with particular kinds of flowers.

Of course, many flowers are wind-pollinated, but they are not as showy as those worked and shaped by animals. Wind-pollinated plants often possess many small, simple flowers rather than a few intricate ones so as to have more target area for the floating pollen to hit.

As a result of such influences, we can see drought-escaping ephemeral wildflowers no bigger than a child's fingernail and others, such as sacred datura or jimsonweed, with six-inch-long, trumpet-like blooms. Cacti can have many miniature bee-pollinated cup-shaped corollas, or a solitary eight-inch tubular perfume factory, such as that of the night-blooming cereus.

We believe that this is among the first wildflower watching books featuring the relationships between pollinators and their plants in the wildflower descriptions. Observe the shapes, sizes and scents of particular wildflowers in your area, and try to predict which animals may be pollinating them. With diligent observations at different times of the day, you may be able to confirm whether your predictions were correct. Since not all bees choose flower colors and shapes according to our general predictions, your observations may add new details to a wildflower's story.

Pollinating birds, bats, bees, and butterflies are not the only animals attracted to native desert wildflowers. Browse through the index of any Southwestern flower book, and you will find plant names like deer nut, coyote gourd, and rabbit brush. These plants provide food, shade, shelter, and nesting sites to many creatures. Although we may be attracted to these plants' showy blooms for a short spurt of time each year, the plants themselves may harbor entire animal communities year around.

(LEFT) *Beavertail prickly-pear have many bee-pollinated, cup-shaped corollas.* KAZ HAGIWARA

(RIGHT) *This butterfly is just one of many important pollinators that influence floral architecture on Arizona's deserts.* JAMES TALLON

Chapter 2
The art of wildflower watching

The equipment needed for wildflower watching seems self-evident: a good pair of eyes and a willingness to drive to and walk through wildlands. Wildflower stalking is a little like breathing—you don't take lessons to learn how, you just do it.

On another level, however, preparedness, practice, finely-tuned senses, and imagination make all the difference in the world. One watcher can see dozens of blossoming species which someone else might glance over and pass by. The fragrance, the dance of pollinators, and the diversity of hues associated with one wildflower might all be overlooked if you are not cued into their existence.

At the risk of explaining the obvious, then, we offer some guidelines for nurturing this art. In part, this is to further encourage the person who has already tried stalking the elusive blooms, but met with mixed success.

After a few forays into the field, frustrations and questions often do arise. Why aren't there flowers out this Easter when there were tons of them last year at this time? Do you have to drive so far? Why do the big stands of blooms seem to be on the highway right-of-way? Can you legally pick them on either side of the fence? And how do you figure out the names of the ones that you don't already know?

To solve such problems, a prepared eye, a few learning tools, a flexible schedule, and a couple of precautions will go a long way. For starters you can benefit from the prepared eyes of veteran wildflower watchers who collaborate in the Arizona Wildflower Network, a public clearinghouse on springtime blooms coordinated by the Desert Botanical Garden in Phoenix. The network offers both a telephone hotline service to all callers and a weekly summary sheet telling where the showiest spring flowering episodes are occurring, based on the most recent observations available.

Here's how the Arizona Wildflower Network functions. Every Wednesday, beginning the first week of March, the Desert Botanical Garden calls the network's "reporters": botanists, amateur naturalists, park rangers, and wilderness managers. Using a list of a hundred or so of the most common spring plants, they report the four showiest species for each accessible locality in their territory, as well as all other plants that have broken out of bud in the previous week. These reports are compiled each Friday of March and April and presented by the hotline's recorded phone message.

The Phoenix phone number for the Arizona Wildflower Network hotline is (602) 941-2867. If you live near a botanical garden, arboretum, or national monument in southern Arizona, you might want to make a local call to one before phoning long distance. They may have the Network's weekly information sheet already in hand.

These same institutions and agencies often sponsor their own wildflower walks with interpretive naturalists leading the way up bloom-filled canyons, or to nearby sites with spectacular floras. Such guided walks are excellent ways to learn about native desert plants.

More and more, the state's botanical gardens, outdoor museums, and arboreta are sowing and showing wildflowers on their own grounds. The Desert Botanical Garden, Tucson Botanical Gardens, Tohono Chul Park in northwest Tucson, the Museum of Northern Arizona, The Arboretum at Flagstaff, Boyce Thompson Southwestern Arboretum, and the Arizona-Sonora Desert Museum feature spring and summer flowering natives along their trails. Before venturing out on a long trip to unknown ground, it may be worthwhile familiarizing yourself with the species growing in the garden nearest your home.

Once you decide to go on the road, you might want to consider a few basic books and equipment to take along as traveling companions. Although this book will get you started on how to find and identify 63 native wildflowers, our reference list suggests some field guides, taxonomic treatments of the Arizona flora, and simple keys for learning to identify plant families and genera.

A hand lens or low-cost magnifying glass can help you sort out the tiny bellyflowers from one another, since most plants are identifiable only through floral characteristics such as petal and anther numbers, corolla and calyx size, and shape of the inflorescence. These often minuscule characteristics seem intimidating at first, until you open yourself up to the beautiful microscopic world of floral patterns.

Photography is another way to enter the plant world. You might want to consider carrying a camera with

(LEFT) *Arizona's botanical gardens, outdoor museums, and arboreta sow and show wildflowers on their own grounds. Before departing on wildflower excursions, photographers can check with the Arizona Wildflower Network hotline at the Desert Botanical Garden to learn where the showiest spring flower episodes are occurring.* DON B. STEVENSON

macro lens or magnifying attachments, a tripod, a small piece of rug or canvas in case you want to go prone next to the mini-flowers, and a piece of black velvet for photo backgrounds.

If you are vulnerable to allergies, you may want to take along antihistamines or a pollen mask for any extended stays in wildflower fields. Although many of the showiest flowers have sticky, heavier pollen suited to dispersal by bees and hummingbirds, wind-pollinated grasses and herbs often grow with them. Certain wildflower stem and leaf juices also cause contact dermatitis, so caution should be used in handling them. Scorpionweeds, euphorbias, agaves, and globemallows may all be considered to be skin irritants.

For those who have permits for plant collecting for scientific or educational purposes, a plant press for making dried herbarium specimens is a requisite for being able to later identify those species too difficult to distinguish in the field. National parks and forests, as well as some Indian reservations or state, county, and city parks, require such permits for any wildflower collecting. In addition, transport after removal of whole plants (for genera such as mariposa-lilies, agaves, and cacti) is prohibited by state law. For a list of protected species, particularly endangered ones, that are protected by state and federal laws, write or call the Arizona Commission of Agriculture and Horticulture, 1688 W. Adams, Phoenix, Arizona 85007; (602) 255-4373.

Although permits can be granted for legitimate collection and study of certain of these rarer or over-exploited plants, caution should be exercised even when merely visiting the areas where they grow. Trampling can destroy unseen seedlings, and inadvertent damage to canopy plants can reduce the protective cover which immature plants need to shelter them from temperature extremes and from predators.

It is legally acceptable to collect native flowers and seeds on private lands with permission of the landowner, but we urge you to minimize such impacts on wild plant populations. Over-harvests can diminish future generations of growth. Why pick naturally-occurring wildflowers when there are so many other ways to enjoy them? A little restraint will allow the next guy to come along to appreciate them as much as you have.

The Desert Botanical Garden in Phoenix sponsors wildflower walks with interpretive naturalists leading the way. Such guided walks are excellent ways to learn about native desert plants. The adobe structure of Webster Auditorium, originally built as the clubhouse of the Arizona Cactus and Native Flora Society, is the focal point for plantings of Sonoran Desert wildflowers. Photographers find plenty of challenges along the paths. PHOTOGRAPHS BY DON B. STEVENSON

Boyce Thompson Southwestern Arboretum, near Superior, features spring and summer flowering natives along hiking trails. A small lake and the varied landscapes of the microclimates within the Arboretum give a good overview of the possibilities for growing wildflowers.

PHOTOGRAPHS BY DON B. STEVENSON

Arizona-Sonora Desert Museum near Tucson is a favorite with visitors around the world. Here they can see flora and fauna native to the region. A combined botanical garden, zoo, and geological center, the museum has been rated one of the top attractions in the state. PHOTOGRAPHS BY DON B. STEVENSON

Tucson Botanical Gardens exhibits many flowers and houses Native Seeds/SEARCH, where volunteers and staff grow and conserve seed, and compile data on native food plants of the Southwest.

PHOTOGRAPHS BY DON B. STEVENSON

Tohono Chul Park in northwest Tucson is a lovely natural history center that features native plants, succulents, cacti, and grasses of the Sonoran and Chihuahuan Deserts.

PHOTOGRAPHS BY DON B. STEVENSON

Chapter 3
Why the desert blooms when it does

There had been a springtime pay-off from the rains which had come in a hurricane fringe storm the previous October. From the flanks of Kitt Peak west of Tucson, on the access road to the National Observatory, we could look out over the valley and see a sunburst of yellow-orange more than half a mile wide.

This display of poppies, though large in size, seemed fragile to us who stood in awe. If a heavy grass growth had occurred the previous fall, the emerging poppies would have been smothered by the standing crop of dead stalks. If no rains had come after late October's gully-washer, most of the seedlings would have died. Unseasonable weather would have scattered or diluted this concentration of color.

Instead, we saw the simultaneous blooming of hundreds of thousands of poppies, a spellbinder. For reasons not clearly understood, such a spectacle would not appear the following year, or the next.

After witnessing such a natural wonder, it is not uncommon for plant enthusiasts to wish to see another. Some desert rats, rather whimsically, attempt to predict when the next will occur. Still, dry land botanists have yet to muster a decent batting average in their predictions of good desert blooms. Plant scientists who feel confident predicting the onset of poinsettias or Easter lilies are humbled by the difficulties of understanding all the natural factors involved in making an awesome wildflower season in Arizona deserts.

Scientists who ponder such matters call their studies *phenology*, meaning "the seasonal timing of life cycle events." The term *phenology* comes from the Greek word *phaino*, "to show" or "to appear." In the following discussion, we offer you a few insights into why showy blooms appear when they do.

Like most desert natural history lessons, this one starts with rain, or the lack of it. Rainfall on Arizona deserts ranges from 4 to 10 inches annually, and adjacent uplands receive 10 to 24 inches per year. Not all of this moisture can be effectively used by plants. Evaporation, most severe during the summer, may account for a five- to eight-foot loss of water per year from an open desert pond. Rainfall is effective in replenishing soil moisture (and plant water storage tissue) during the

times of the year when it exceeds evaporation. Such moments are fleeting.

The brief periods when rainfall exceeds evaporation and temperature limitations are also likely to be those when short-lived wildflowers germinate and establish a seedling stage. If you look at climatic diagrams for southern Arizona (Figure 1), you can easily target when such periods occur *on the average*.

In general, both warm season and cool season wildflowers need drenching rains of an inch or more to trigger their germination on desert soils. A heavy rain saturates the upper soil, unburies seeds and exposes them to light, washes them to new sites, abrades their seedcoats, and deposits needed nutrients on the soil surface—all at once. On poorer, heavier soils, an additional quarter of

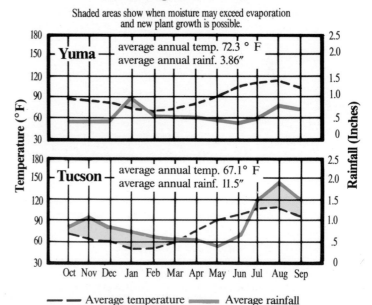

Comparison of rainfall patterns in the Sonoran Desert and resulting wildflower shows

Shaded areas show when moisture may exceed evaporation and new plant growth is possible.

Yuma — average annual temp. 72.3 ° F
average annual rainf. 3.86"

Tucson — average annual temp. 67.1° F
average annual rainf. 11.5"

Oct Nov Dec Jan Feb Mar Apr May Jun Jul Aug Sep

— — Average temperature ▬▬ Average rainfall

Figure 1: These monthly averages show the brief periods during the year when soil moisture may exceed evaporation. Yuma has an unpredictable rainfall pattern and averages only a short period in the winter when moisture amounts may exceed evaporation. Tucson, with both a winter and a summer rainy season, gives short-lived wildflowers more opportunity to establish seedlings. Shaded areas indicate when new plant growth is possible.

(LEFT) *The sight of the simultaneous blooming of hundreds of thousands of goldpoppies can be a spellbinder.* DAVID MUENCH

an inch may be needed to generate the same blooms that an inch alone can bring to rich soils with good water-holding capacity.

But in certain instances, if earlier storms have already moistened the soil crust, lighter rains of a quarter to half inch are all that are needed to stimulate runoff and germination on any kind of soil. Still, the vast majority of wildflower seedlings arise from just a few dazzling downpours of an inch or more.

And yet, rainfall is not all that is required to stir desert seeds from their sleep in the soil. Thousands of seeds per cubic foot of ground may stay dormant after a heavy rain if other factors such as temperature, light intensity, day length, and soil salinity are not favorable. That is why spring wildflowers seldom emerge as seedlings during mid-summer, and summer wildflowers will not germinate with rains in November.

Rainfall on Arizona deserts ranges from 4 to 10 inches annually with adjacent "sky island" uplands receiving 10 to 24 inches each year. Tucson, seen here from the top of the Catalina Mountains, has both a winter and a summer rainy season. JACK DYKINGA

Germination of cool-season wildflowers may be triggered by a rain six or seven months before they bloom. Early in that interval, most of us hardly notice that new plants have emerged. They are so tiny and grow so slowly that we overlook them. For this reason, many desert dwellers believe that rains or snows around Christmas initiate Easter blooms in the desert. Yet by winter solstice, December 21, the cool season herbs that will flower later have already emerged, and few others are recruited, if any at all. In general, September-October rains influence spring wildflower abundance more than any other factor.

Cool season wildflowers not only require sufficient rain, but need to pass through certain temperature conditions in order to germinate. Seeds of chias and desert sunflowers often end up on or near desert soil surfaces which reach 120 degrees Fahrenheit for several weeks each year. The seeds have come to require such heat to reach full maturity and germinability. Likewise, there is a lower "window of temperatures" within which they will germinate. Even if rains reach fully mature seeds of cool season wildflowers during the summer, they will remain dormant until temperatures drop below 100 degrees.

The same late September to early December rains which trigger ephemeral herbs also prepare desert trees and shrubs for successful growth and reproduction during the following winter and spring. At the Mohave Desert sites near the Nevada-Arizona border, key rains from late September to early November are infrequent, but cause mass germinations. Rains from mid-November to January are more reliable but only a scatter of seedlings results. About one in four years, such a fall or early winter storm fails to arrive at all.

As you travel eastward, from the Mohave Desert to the Chihuahuan Desert, the winter rains make up a smaller portion of total annual rainfall (Figure 2). The initiation of this cool winter season begins later and later, and its intensity diminishes as you move from west to east through the desert.

The central Sonoran Desert does not have as showy a cool season as the Mohave does, and the Chihuahuan region may altogether lack enough winter rain for any spring bloom at all. Yet what the Sonoran and Chihuahuan lack in winter, they make up for in their warm season bloom, which is triggered by the summer monsoons. In Tucson, the summer rainy season has more dramatic thunderstorms and wildflower shows than in Phoenix (Figure 3). Arizonans receive the best of both worlds since the Sonoran Desert, which straddles much of the state, receives a little of both seasons of color.

The summer rainy season's cover of herbs is just as unpredictable as that of the winter crop. The needed rains may take place anytime from June 29 to August 10 in the Sonoran Desert/Upland edge near Sells in Southern Arizona. At Sells, summer storms of an inch or more may arrive as late as August one in every six years.

It may take at least another inch-deep rain during the winter, and two inches or more spread over late winter and early spring, to allow large stands of early spring flowers to survive in vegetative form. Most of these annuals won't produce floral buds until days begin to lengthen and higher temperatures set in. Some studies suggest that they must be stressed by limited moisture for flowering to be triggered.

Even if minimal conditions are met, other factors may prevent a spectacular spring wildflower event. A dense cover of dead plants from previous seasons may suppress a full display of spring bloomers. And since severe frosts damage the blossoms of some species, the timing of freezes affects the intensity of a flowering landscape.

Warm season wildflowers are of at least two sorts. Some are "day-neutral," meaning that they may flower anytime from March to October, whether periods of daylight are 10 or 14 hours long. Others have their flowering clocks set by the shortening of daylight periods in late summer or early fall. They cannot bloom around summer solstice even if moisture and temperature ranges are favorable.

Much of what makes plant populations showy or not is their patterns of flowering. Botanists describe such patterns in terms of the earliest and average opening of a plant's first flower, the range of time over which a single plant flowers, and the degree of synchrony between one plant's bloom and that of another. The key to casting color onto a landscape is the degree of synchronous display of flowers for all plants in the population.

Other species may have a longer flower time, but if each of their individuals flowers on a different day, the results are not as spectacular as those for a species which briefly blooms but all at the same time. In fact, mass flowering is more common among short-duration bloomers such as poppies than for those such as sacred daturas or mallows which produce a few flowers per day over a long duration.

Text continued on page 43

Figure 2: The cool winter season begins later and its intensity diminishes as you move from the Mohave Desert in the west to the Sonoran and Chihuahuan deserts farther east. Also, as you move from west to east, winter rains make up a smaller percentage of total annual rainfall and summer rains dominate.

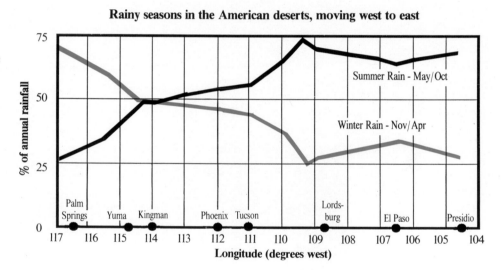

Rainy seasons in the American deserts, moving west to east

Figure 3: Our expectations for short-lived "ephemerals" are linked to rainfall at the end of the previous season. The rainfall in late summer in Tucson assures an ephemeral flower show later that fall. Strong winter rains may cause a good show of ephemeral and warm season perennial blooms the following spring. Overlapping bloom times for perennials, legumes, and cacti promise a maximum wildflower show. In good years Phoenix' and Tucson's summer and winter rainfall assure more wildflower shows.

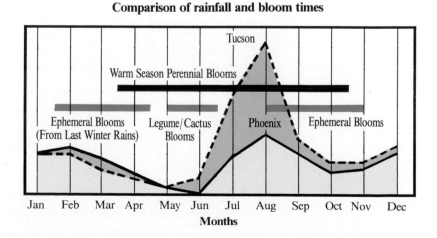

Comparison of rainfall and bloom times

The monsoon season in Arizona usually arrives about mid-July, bringing spectacular sunsets to Saguaro National Monument near Tucson. WILLARD CLAY

Text continued from page 39

This same phenomenon may hold true for desert trees and shrubs. Foothills paloverdes light up the desert when they bloom, but their entire season may run less than three weeks. Desert-willow flowers may open and fade over a seven-month period, so they seldom change our view of the landscape so dramatically.

The flowers of trees and shrubs are generally not as tied, as are annuals, to brief periods following particular spurts of rain. Some of the woody perennials are deep rooted enough to tap underground reservoirs of water unavailable to shallow-rooted herbs.

And yet, trees such as paloverde may still have a bloom period of just a few weeks. Individuals of the same species are often triggered to bud at the same time, but blooming then tails off on a different day for each plant. A short bloom period may indicate dependency upon particular migrating bird, bat, or insect pollinators, without which these plants produce fewer fertile seeds. In essence, a woody species which has all of its individuals

(LEFT) The Arizona Uplands type is the best known desert vegetation in Arizona. The saguaro is an impressive example of this vegetation. JERRY JACKA

(BELOW) This queen butterfly on a tomatillo bush is a winter visitor to Saguaro National Monument in southern Arizona. JAMES TALLON

blooming simultaneously is more easily noticed by pollinators that will shift to those flowers to gain the additional rewards.

In Arizona, ocotillo may bloom anytime between March 4 and late May, but those more to the south or at lower elevations generally flower earlier than those to the north, or in higher mountains. The bloom of each population usually follows by a few days the peak time of arrival of migrating hummingbirds.

However, if cold weather holds back the hummingbirds, but the ocotillo blooms "on time" anyway, seed set for that year can be extremely low.

You aren't the only one who suffers if the wildflowers don't happen to bloom during your spring vacation week. The plants that need to attract migrating pollinators may not reproduce as well if their blooms fail to attract these animals. The animals themselves may lose weight for lack of nectar and reproduce less if they reach a place before the flowers open. Birds may have several thousands of years of a headstart on us in learning the timing of flowering. Yet their timing, too, can go awry as easily as ours.

Regardless of the gains science has made in recent years in understanding the relationships between plants, their climate, and pollinators, wildflower watching remains a gamble. But if you play your cards right, you may reach the gold of a million poppies at the end of a rainbow.

(ABOVE) *Of the 16 species of hummingbirds that breed in the U.S., 13 are found in Arizona. They are colorful and popular pollinators.* JAMES TALLON

(RIGHT) *The bright, orange-red tubular flowers of the ocotillo occur in dense clusters at the tips of long stems. Each stand may offer birds nectar for only one or two weeks, but many hummingbirds are attracted during that brief period.* JAMES TALLON

(BELOW) *The blue-throated hummingbird of south-eastern Arizona takes both nectar and insects. It will return to a favorite site near flowering vegetation and water year after year.* G. C. KELLEY

Wildflower regions

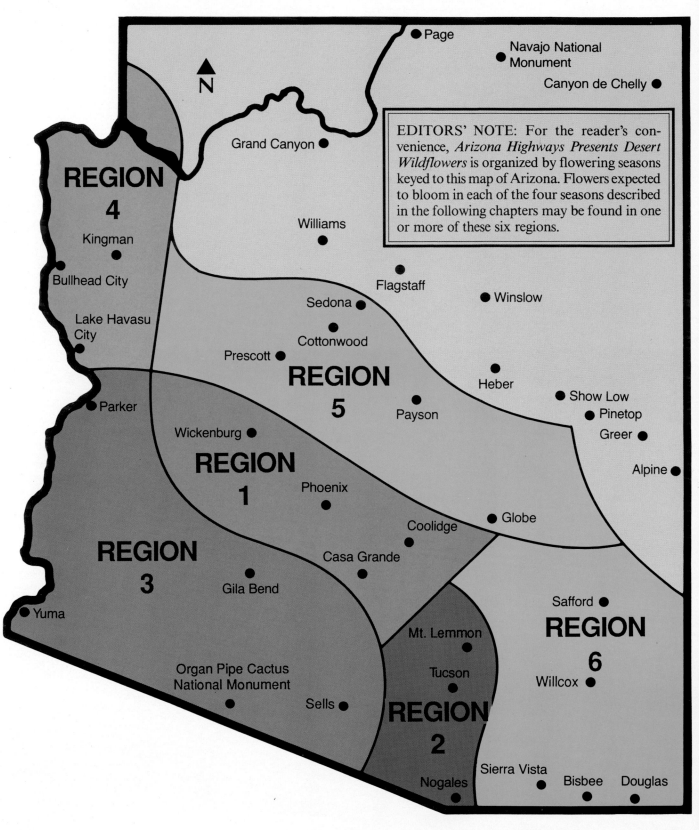

Page

Navajo National
Monument

Canyon de Chelly ●

EDITORS' NOTE: For the reader's convenience, *Arizona Highways Presents Desert Wildflowers* is organized by flowering seasons keyed to this map of Arizona. Flowers expected to bloom in each of the four seasons described in the following chapters may be found in one or more of these six regions.

N

Grand Canyon ●

REGION 4

Kingman ●

Bullhead City ●

Lake Havasu City ●

Williams ●

Flagstaff ●

Winslow ●

Sedona ●

Cottonwood ●

Prescott ●

REGION 5

Heber ●

Show Low ●

Pinetop ●

Greer ●

Parker ●

Wickenburg ●

Payson ●

Alpine ●

REGION 1

Phoenix ●

Globe ●

Coolidge ●

REGION 3

Casa Grande ●

Gila Bend ●

Safford ●

Yuma ●

Mt. Lemmon ●

REGION 6

Willcox ●

Organ Pipe Cactus National Monument ●

Tucson ●

Sells ●

REGION 2

Sierra Vista ●

Nogales ●

Bisbee ●

Douglas ●

Adapted from Arizona Climate, 1974

Chapter 4
Late winter/early spring wildflowers

The flowering season on southwestern deserts in late winter through early spring does not occur in other parts of North America. While the Midwest is still adrift in snow, the Sonoran and Mohave Deserts are breaking loose into a dozen different colors. These plants begin to bloom when days are still short, less than 10 or 11 hours, and when light frosts are not uncommon. The blossoms of many of these species can tolerate quite heavy freezes of several hours duration. In contrast, flowers, buds, and flowering stalks of plants such as ocotillo and agave may be damaged if they suffer the freezes of late winter.

Most of the winter bloomers in Arizona are also found in the Mohave Desert sections of California and Nevada. These plants germinate in mass as early as September or October if the tropical hurricane fringe storms from the Sea of Cortez offer the first heavy rains in the fall season. If an inch of rain arrives before mid-December, there still may be scattered germination. Some require a period of dormancy through the desert heat to after-ripen, and then need cooler ground temperatures to germinate.

A common growth form for these ephemeral (short-lived annual) species is that of a low-lying rosette of simple, undivided leaves. A ring of ground-hugging leaves emerges, growing slowly over the winter months. The leaves angle to track the sun, and several species have reddish (anthocyanin) pigments in them. Both of these features allow them to heat up considerably, even on the coldest days.

As the new year leaves the short days of winter solstice behind, the lengthening period of daylight triggers these plants to make floral buds. Then, when the ground surface of the desert begins to heat up, these plants bolt, much as lettuce or spinach does, putting out an elongated flowering stalk. For most winter ephemerals, the flower stalk is "indeterminate" or open-ended, meaning that it can continue to produce flowers and seeds as long as moisture is available and air temperatures are not high enough to damage these plants' tissue. In a cool, wet year, or when gardeners artificially water them, they will continue to flower into April. However, most winter ephemerals peak before the spring equinox in late March. They generally have more above-ground growth relative to their root sizes than do warm season herbs.

The winter ephemeral wildflowers most frequently seen at the onset of the season include yellow cups and other evening-primroses; bladderpod mustards; loco-weeds; the non-showy Indian-wheats or plantagos; blue-dicks or Papago-lilies and the introduced heronsbill. On sandy soils, the herbaceous perennial ajo-lily is the show-stopper during this season.

Shrubs such as jojoba, fairyduster, and brittlebush may begin to bloom anytime after winter solstice. Some "binarial" or bimodal wildflowers (plants which bloom in two seasons), such as sand-verbena and snowy sun-flowers, also get an early start. In warm winters, particularly in Regions 3 and 4 in the western part of the state, lupines and poppies may begin to flower in late January. Yet these and other flowers still reach their peak after spring equinox in most localities.

As one might expect, the winter wildflowers are largely phenomena of the warmest areas of the state, Regions 3 and 4. These regions are also more prone to early winter rains, if they receive any at all. Regions 1 and 2 harbor some of the same species, but flowering there peaks much closer to spring equinox, so that this season seems to fuse with the following season of early spring bellyflowers.

To characterize the blooming variation of winter-initiated wildflowers, we offer graphs of three representative species: the ajo-lily, an herbaceous perennial; lupine, an herbaceous annual; and fairyduster, a woody perennial. These records were taken on plants at the Desert Botanical Garden, under minimal irrigation, from 1978 through 1986. The Phoenix bloom times are earlier than those from Tucson, reported from 20 years of observation by Dr. William McGinnies, but probably later than those for Yuma or other sites in Regions 3 and 4.

We caution you against the use of the dates in these graphs as absolute guides to finding the flowering times of plants in your particular locality. When the winter-spring season really occurs in one region versus another depends on killing frosts, daytime temperature highs, and other weather factors that shape the duration of the season differently from place to place.

Typical representative flowers of the late winter/early spring bloom season

ajo-lily
herbaceous perennial

Records of ajo-lily bloom times, recorded at the Desert Botanical Garden in Phoenix for eight years, show the plant blooming most frequently from the beginning of March to the beginning of April. In some years plants bloomed as early as the first two weeks of February and as late as the middle of April. (See page 52 for plant description.)

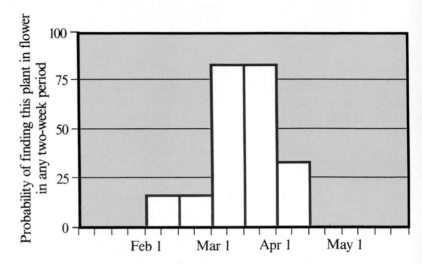

lupine
herbaceous annual

Lupines, blooming in plantings at the Desert Botanical Garden, may appear any time from the first week in February to the middle of May. Records there show that March and April are the most likely months for the flowers to appear. (See page 54 for plant description.)

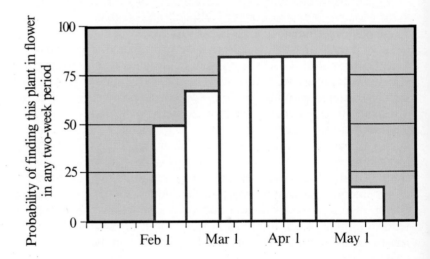

fairyduster
woody perennial

Fairyduster has long periods of bloom on individual plants. Major bloom time for this small shrub at the Desert Botanical Garden is from the end of February to the beginning of April. Based on these records, an earlier bloom time in November and December is also shown. (See page 50 for plant description.)

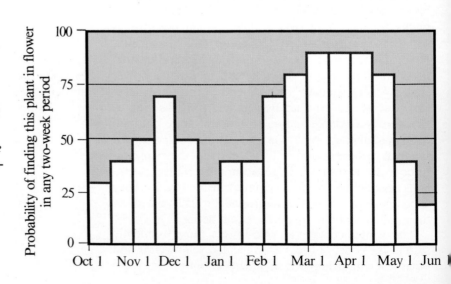

Brittlebush, common throughout large areas of the Sonoran Desert, appears in large patches on the rocky foothills of the Superstition Mountains. JERRY JACKA

KATHLEEN NORRIS COOK

sand-verbena
Abronia villosa
desert sand-verbena
Four O'Clock Family

Sand-verbena has stalked heads of pink-purple to rose-colored "flowers" or colored bracts. Sand often sticks to its thick leaves in the dunes and desert washes where it grows. It may bloom any time from the first of February to early October. There is usually one peak in flowering around mid-February, and if soil moisture persists, another peak may occur any time between May and late August. In spring's mixed stands of wildflowers, it adds greatly to the palette of colors.

Regions 2, 3

JAMES TALLON

desert-marigold
Baileya multiradiata
paperdaisy, woolly-marigold,
Sunflower Family
hierba amarilla, baileya del desierto

Blooming in spurts over a long period, this desert herb is also the showiest common member of the aster and sunflower group. The omnipresence of the desert-marigold along roadsides and its long woolly stalks and lovely lemon hued flowers make this plant unforgettable. To confirm its identity, check to see if the seeds have withered ray flowers attached to them. Its flowering flush from early March to May is best known, but it may also contribute as much as a third of the August pollen harvest by desert honeybees, indicating another late peak. Where runoff accumulates, desert-marigolds bloom almost year around throughout the desert and into some grassland habitats.

Regions 1, 2, 3, 4, 5, 6

fairyduster
Calliandra eriophylla
false mesquite, mesquitillo,
Bean Family
cosahui, tabardillo

This small shrub has clustered rose colored flowers with numerous stamens that protrude like brushes beyond the length of the blossoms themselves. The stamens hold packages of pollen called polyads, each with a sticky place for attaching onto pollinators. These flowers are set in lacy, acacia-like foliage, with five to 15 pairs of leaflets. Flowering from October through April and peaking near the end of March, fairydusters can be found from rocky hillsides and canyon walls to the banks of arroyos. Butterflies, bees, and perhaps some humming-birds are its pollinatiors. They are often the only winter-blooming shrubs on desert mountain slopes.

Regions 1, 2, 3, 4, 5, 6

JIM HONCOOP

evening-primrose
yellow cups, hierba del golpe

Camissonia brevipes
[Oenothera brevipes]
Evening-Primrose Family

This desert dweller is not an evening bloomer as its name suggests. Its yellow corollas open early in the morning when large bees and sphinx moths visit them. They become red-tinged upon pollination and closure. Yellow cups is one of several related annuals whose delicate flowers are on branches rising from a heavy taproot. Its petals reach half an inch in length while the stamens are only a quarter-inch long. The stigma is club-shaped rather than four-lobed like most other evening-primroses. After flowering, stalked, linear seed capsules develop, dry, and open. Yellow cups blooms in Arizona from February 15 to the start of May but peaks in early March. It can be found at low elevations on desert washes and plains.

Regions 3, 4

JOSEF MUENCH

pincushion plant
Esteve's pincushion

Chaenactis stevioides
Sunflower Family

This annual differs from the related Fremont's pincushion by its cream or yellow (rather than white) flowers and by having hairy, flattened leaves rather than glabrous, fleshy ones. It is a common component of wildflower mixtures on desert plains and broad washes, blooming from mid-February to the end of April. Its densities can vary fiftyfold on the same spot, between wet and dry years. Although short-statured, pincushion plants sprinkle much color into desert grassland and chaparral openings, peaking around early March in the low desert and later at higher elevations.

Regions 1, 2, 3, 4, 5

JUDY L. MIELKE

brittlebush
incienso, hierba del vaso,
rama blanca, hierba del gusano

Encelia farinosa
Sunflower Family

Brittlebush forms the most extensive late winter flower show in Arizona. In frost-free areas, blossoming begins in November, and sometimes persists to mid-May. Flowering is most intense in March and April. Many insects are attracted to its long-stalked, sunflower-like bouquets. On volcanic hillsides, the gold contrasts vividly with the black rock. Brittlebush foliage changes radically, depending upon seasonal moisture. The first new ones after a rain will be larger and greener, but as the soil dries, greyish leaves a hundred times smaller in area will be produced. These latter leaves will persist under mild drought, whereas the first ones will be dropped to reduce water loss. A gum exuded from the stems can be burned as an aromatic incense, hence the Spanish name *incienso*.

Regions 1, 2, 3, 4

DAVID MUENCH

CHARLES MANN

blanketflower
Arizona blanketflower

Gaillardia arizonica
Sunflower Family

Unlike the widely-cultivated ornamental gaillardias pictured here, *Gaillardia arizonica* has an all-yellow disk and ray flowers that lack the purplish tinges found in other species. A variable spring annual, it has leaves that are either deeply divided or merely toothed, along with solitary flowers on stalks with or without alternate leaves. It blooms as early as February and March at lower elevations in the desert and as late as July in desert grasslands and chaparral up to 4,000 feet.
Regions 1, 2, 3, 4, 5, 6

CHARLES MANN

Goodding's verbena
verbena, southwestern vervain

Glandularia gooddingii
[Verbena gooddingii]
Vervain Family

The mauve or lavender flowers of verbena form large headlike clusters atop hairy perennial foliage. Enough of these plants emerge on the same bajadas and mesas to produce a carpet-like effect. Goodding's verbena can be found throughout Arizona below 5,000 feet. Forming large patches, it attracts many butterflies and moths. It blooms from February through October but peaks in late March.
Regions 1, 2, 3, 4, 5, 6

RICK ODELL

ajo-lily
desert-lily, ajo silvestre

Hesperocallis undulata
Lily Family

The large white trumpets of this sand-loving plant resemble those of the Easter lily. Its edible bulb is reminiscent of garlic, *ajo* in Spanish. The flowers emerge from a rosette of ribbony leaves, sometimes at ground level, other times on a short stalk. The sphinx-moth pollinator is attracted to its deep spurs filled with fragrant nectar. Although most common and larger in moisture-laden dune areas, this lily also inhabits gravel flats below 2,000 feet. Its bloom period reaches from mid-February to mid-April, peaking in late March.
Regions 1, 2, 3

chuparosa
chuparroso, hummingbird bush, beloperone

Justicia californica
Acanthus Family

This sprawling, pale-stemmed shrub produces enough red flowers early in the spring to cause territorial battles among Anna's, Costa's, and black-chinned humming-birds. The tubular flowers bloom profusely, and tide most hummers over in the winter until other plants begin to bloom. Frequenting the sand and gravel washes skirting low desert mountain ranges, chuparosa always grows below 2,500 feet. Its bloom time extends from late August through June, depending upon the locality.

Regions 1, 2, 3

RICHARD SCHNITKEY

creosotebush
greasewood, gobernadora, hediondilla

Larrea tridentata
Caltrop Family

These strong-scented, resinous shrubs form extensive stands throughout the warm deserts of the Americas. Although the yellow, five-petalled flowers are small, they offer color where drought eliminates other hues. The 5,000–50,000 flowers produced by a single creosotebush in an average year attract bees, wasps, and flies. Petals twist 90 degrees if they've been pollinated. Sonoran Desert creosote blooms peak in March and April, and again in November through December. Flower bud enlargement is triggered by a drenching rain following a drought, but flowering itself does not occur until the soil begins to dry out again. The bush is found on flats, bajadas, and hills below 4,500 feet.

Regions 1, 2, 3, 4, 6

JAMES TALLON

bladderpod mustard
Gordon's bladderpod

Lesquerella gordoni
Mustard Family

A golden-flowered miniature, this early annual forms enormous stands by itself during years when rainfall is so low that other flowers fail. It prefers plains and open valleys of the deserts and adjacent grasslands. Bladder-pod extends into other habitats as well. Several species of these mustards have been evaluated as desert crops be-cause of their unusual vegetable oil quality and potential industrial uses. They bloom and are bee-pollinated from mid-February to mid-April. Highly responsive to rain-fall, their population's pollen production can increase by a factor of two hundred if winter precipitation is tripled from one year to the next.

Regions 2, 3, 6

JERRY JACKA

JERRY SIEVE

lupine
Coulter's lupine

Lupinus sparsiflorus
Bean Family

This annual lupine species has violet-blue, pea-like blooms with yellow spots on one petal that turns purplish-red when the flower is manipulated by bees. Lupines have finger-like leaflets which tilt to track the sun at a direct angle, thereby gaining additional solar radiation during the winter when such energy is at a premium. Some years, lupine begins to flower in January. Attractive to bumble bees and digger bees, these flowers peak in color and fragrance from mid-March through mid-April, on alluvial fans, washes, and canyons below 2,300 feet.

Regions 1, 2, 3, 4, 6

GARY P. NABHAN

scorpionweed
wild heliotrope,
caterpillar weed

Phacelia crenulata
Waterleaf Family

Violet-purple, on rare occasions white, these flowers form on one side of a curled branch in a manner suggestive of a scorpion's tail. Scorpionweeds are sticky annual plants with ill-scented, highly divided oblong leaves. They are bee-pollinated, but their pollen productivity in a flower population varies fiftyfold from year to year. Likewise, the duration of their flowering season may range from a week in a dry year to eight weeks in a wet one. Peak bloom time varies from late March to late April, when it is extremely common from grassland mesas down into low deserts. A related species, *Phacelia bombycina*, replaces this one in southwestern Arizona.

Regions 1, 3, 4, 5   *Phacelia crenulata*
Regions 2, 6   *Phacelia bombycina*

JUDY L. MIELKE

LON McADAM

jojoba
goatnut, coffee berry

Simmondsia chinensis
Goatnut Family

Although this desert shrub lacks showy flowers, its floral clusters and waxy fruit still intrigue many people. The plant shape guides wind-dispersed pollen toward female flowers, but bees also steal pollen from male plants. Jojoba prefers rocky slopes with coarse soils, above pockets of cold air drainage. The liquid wax extracted from its seeds is currently marketed as a key ingredient in many cosmetics. Jojoba has a long and variable flowering season. At one site, it has extended from mid-January to late-April in some years, but only February and early March in others.

Regions 1, 2, 3, 6

globemallow

desert-mallow, hollyhock,
sore-eye-poppy, mal de ojo

Sphaeralcea ambigua
Mallow Family

JAMES TALLON

The large number of orange-colored flowers produced by this multi-stemmed mallow over a year provides a steady source of pollen and nectar to honeybees and mallow-specializing bees. Flower color variants occur, including ones with white, purple, red, or grenadine hues. A low-growing perennial herb, desert mallow has long panicles of flowers and roundish, shallow-lobed leaves. The leaves have star-shaped hairs on them that irritate the eyes if accidentally rubbed into them. From dry rocky slopes to washes and the banks of springs, this is one of the most adaptable of the mallow species, which seldom exhibit so much drought tolerance. Below 3,500 feet, this species blooms year around, although each plant has an individual bloom time.

Regions 1, 2, 3, 4, 5, 6

(BELOW) *Grizzly bear cactus and desert-marigolds add color to the Virgin Mountains foothills in the north-west corner of Arizona.* DAVID MUENCH

DAVID MUENCH

Chapter 5
Mid-spring wildflowers

Mid-spring offers the greatest diversity of ephemeral (short-lived annual) wildflowers of any season occuring in Arizona. It is also the best known of our flowering seasons, since many Eastertime visitors take desert drives or hikes between late March and mid-April when these flowers are peaking.

This season begins by spring equinox in Regions 1 to 4 in the state, since the last killing frost of the cool season has passed by that time. Although these plants grow during cool, sometimes below-zero weather, most of their delicate flowers open after such threats have passed. The nights remain cool, but the sunny days are warm enough to bring out solitary bees that do their pollinating after freezing weather ends. March-April wildflowers generally need the longer days to trigger their clocks to begin flowering, but all will begin before May when saguaro season arrives.

Bellyflowers, or miniature ephemerals, now are in contrast to the later spring season dominated by woody perennial legumes and cacti. In addition to the Mohavean elements derived from California and Nevada that crop up in the late winter, the mid-spring includes many widespread annuals that bloom as "summer wildflowers" in cooler, non-desert areas. Rosette plants with single flowering stalks are still common, but large, multi-headed clusters such as verbenas also appear.

The earliest yuccas (Joshua-tree) and cacti (hedgehog) are among the perennials that flower before the first 100-degree days. Yet most of the other spring bloomers do not have the metabolism for drought or heat tolerance, or use water as efficiently as these succulent plants do.

Several plant families dominate this flowering season: the sunflowers, herbaceous beans, borages, and water-leafs, plus introduced mustards. The sunflower family is represented by Esteve's pincushion, desert chicory, desert daisy, desert sunflower, paperflower, goldfields, and tidytips. Herbaceous beans include lupines, lotuses, and the wild mitten-leaf bean. Poppies, chias, and owl-clovers, as well as the year-round blooming desert-marigolds, make a roadside rainbow early in this season.

Regions 1, 2, 3, and 4 present the greatest show during mid-spring. At the same time, Region 5 remains cold enough that only small flowers such as namas, drabas, and lotuses can be seen.

As representatives of the variation in flowering times found from year to year during this season, we have chosen the following plants: poppies, which are annuals; beardtongue, which is a perennial; and Joshua-tree, a woody perennial. These three species have blooming periods which sometimes extend into the adjacent seasons, but they are always centered in the few weeks following the spring equinox.

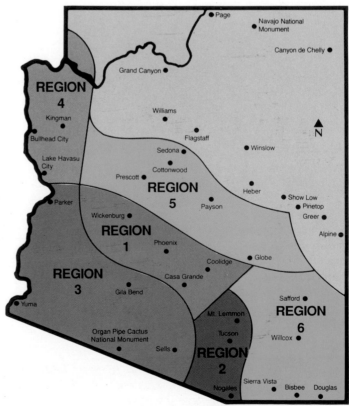

(LEFT) *When poppies and lupines are in bloom, Picacho Peak State Park is a very popular place for wildflower watchers.* JERRY JACKA

Typical representative flowers of the mid-spring bloom season

poppy
herbaceous annual

Individual poppy plants bloom simultaneously in a stand, often beginning in mid-February in Phoenix. Desert Botanical Garden records show March through May as the most likely time to find goldpoppies in bloom among the Garden plantings. (See page 61 for plant description.)

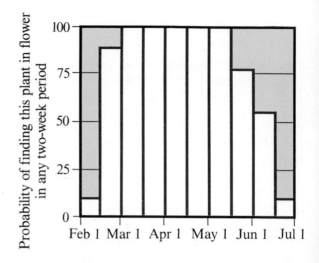

beardtongue
herbaceous perennial

Beardtongue may bloom in mid-February but, in most years, puts on its showiest flowering time through March and the beginning of May. (See page 64 for plant description.)

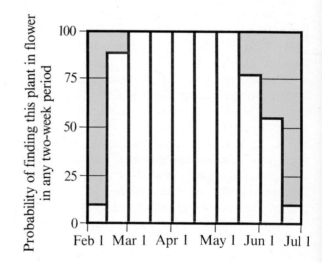

Joshua-tree
woody perennial

Simultaneous blooms of Joshua-trees light up the Mohave Desert in its native habitat. Records at the Desert Botanical Garden show that it has a short bloom time: the last two weeks of March in most years. (See page 65 for plant description.)

(FAR RIGHT) *The ocotillo is common to the Chihuahuan, Mohave, and Sonoran deserts. This one was photographed in northwestern Arizona.* JOSEF MUENCH

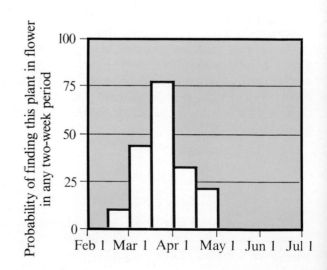

INGE MARTIN

mariposa-lily
Calochortus kennedyi
desert mariposa, sego-lily   Lily Family

This striking spring bloomer can be differentiated from other mariposas by its deep orange-red to yellow or vermilion velvety petals, each with a brownish purple spot at the base. In some places, mariposas are beetle-pollinated. The desert mariposa has clusters of two to four flowers on a stalk rising from a bulb hidden beneath linear, basal leaves. In Arizona, its blossoming is largely limited to April, peaking in mid-month. It is found in many habitats, from desert to grassland and semi-arid woodland vegetation.

Regions 1, 2, 3, 4, 5, 6

WILLIS PETERSON

larkspur
Delphinium scaposum
wild delphinium,   Buttercup Family
barestem larkspur, espuelita

This gorgeous larkspur emerges on a leafless stalk arising from a basal cluster of angled leaf blades and a woody rootstalk. Its sepals are an intense royal blue, its two-toothed spur bronze or purple, and its true petals are blue or white with bluish tips. Larkspurs occur in sizable colonies, where bumble bees and hummingbirds may fight each other for control of nectar-rich territories. They flower March to May on mesas, knolls, and hillsides, from low deserts up to 7,000 feet in the pine zone.

Regions 1, 2, 3, 4, 5, 6

strawberry hedgehog
Echinocereus engelmannii
hedgehog cactus, strawberry cactus,   Cactus Family
torch cactus, Engelmann's cactus

The reddish-purple or magenta flowers of hedgehogs rupture the skin of this cactus just above the areoles, the places where from two to six central spines cluster along the plant's ribs. The curved spines of this cactus are variable in color, from white or gold to black. Its cup-shaped flowers open for several consecutive days, attracting bees and beetles to their abundant pollen and nectar. Medium-sized bees alight on the stigma and probe into the masses of pollen below. As a bee collects one flower's pollen, it leaves behind other pollen, aiding cross-pollination. Hedgehog blooms often last only two weeks, beginning at the end of March or as late as mid-April. It grows on outwash fans, flats, and hillsides from sea level to 5,000 feet in grasslands.

Regions 1, 2, 3, 4, 5

J. PETER MORTIMER

ocotillo
Fouquieria splendens
albarda, candlewood, coachwhip   Ocotillo Family

The five waxy petals of ocotillo form a nectar-laden tube on its wand-like, thorny stems. Ocotillos range widely across rocky slopes in grasslands and deserts below 4,500 feet. Usually synchronous with the spring migration of hummingbirds, ocotillo populations bloom from south to north and from low to high elevations. Each stand offers birds nectar for three to six weeks. Statewide, the blooms extend from early March through late May. Their scarlet tubes attract carpenter bees as well as the following hummingbirds: Anna's, black-chinned, broad-billed, broad-tailed, Costa's, and rufous. After a good rain, their leaves will emerge and the green stems expand. With drought, the leaves drop to reduce water loss. Ocotillos are occasionally planted as living fences.

Regions 1, 2, 3, 4, 5, 6

JIM HONCOOP

JOSEF MUENCH

poppy
Eschscholtzia mexicana
Mexican goldpoppy, copa del oro,   Poppy Family
goldpoppy, amapola del campo

Easily the most popular, most photographed annual wildflower in the Sonoran Desert, this poppy's petals are usually gold to orange in color. Sometimes white and pink mutants show up amidst large golden stands. Its highly dissected, lacy foliage gives rise to solitary-stalked, four-petalled blooms on plains, bajadas, and mountain slopes below 4,500 feet. Probably better pollinated, its blooming begins in mid-February in warmer locales like Phoenix and Yuma but is more limited to March and April near Tucson and Tumacacori.

Regions 1, 2, 3, 4, 5, 6

DAVID MUENCH

desert-sunflower
Geraea canescens
desertgold, desert sunshine,   Sunflower Family
hairy-headed-sunflower

This fuzzy-leaved, hairy-headed–sunflower reaches heights of two feet in gravelly washes and sandy flats within lower desert areas. Its large, yellow radiate heads are both showy and aromatic, attracting both bees and hummingbird-moths. Bees that gather its nectar daily also use evening-primrose pollen. Desert sunflowers tolerate some disturbance and active sand movement in dune fields found below 3,000 feet in elevation. Flowering is most obvious in early April but may extend from January to July when blooms in different localities are compiled.

Regions 1, 2, 3, 4

JUDY L. MIELKE

DESERT BOTANICAL GARDEN

gilia
starflower

Gilia latifolia
Phlox Family

This pink-lavender, funnelform flower is one of many gilias within Arizona. It is an annual that is frequently visited by hummingbirds and butterflies. In wet springs, it appears in large stands on sandy soils in the Colorado and Gila river valleys, always below 2,000 feet. It may be found blooming anytime from mid-March to late May.
Regions 3, 4

BILL CRAMER

goldfields
baeria

Lasthenia chrysostoma
Sunflower Family

This low-growing, tiny-flowered annual has linear, opposite leaves and forms small clumps or large carpets after good winter rains. Its small yellow, terminal heads seldom reach six inches in height. Found on mesas and plains from 1,500 to 4,500 feet in elevation, goldfields can carpet sandy flats for weeks. They bloom as early as February 20 and as late as mid-April but peak in the middle of March.
Regions 1, 2, 3, 6

JUDY L. MIELKE

tidytips
white tidytips

Layia glandulosa
Sunflower Family

This annual member of the sunflower clan has pure white ray flowers and yellow disks. It is a low-growing resinous herb, with hairy divided lower leaves. Although seldom abundant, it is found mixed in with other wildflowers on dry slopes, open grassy valley floors, and mesa tops. Blooming mid-February to late April, it peaks in late March.
Regions 1, 2, 3, 4

blackfoot daisy

desert daisy   *Melampodium leucanthum*
Sunflower Family

A dependably blooming daisy with white, purple-veined ray flowers, this species is a perennial herb or subshrub. Its many flower heads range from a little more than an inch in radius to about a third that size. Whatever it lacks in flower size it makes up in persistence, with some flowering from mid-March through to December. On drier, limestone slopes in southern Arizona, however, flowering peaks in late March and doesn't last past September. Desert daisies can be found in desert grassland and oak woodland between 2,000 and 5,000 feet.

Regions 1, 2, 4, 5, 6

JERRY JACKA

blazingstar

stickleaf, pega ropa,   *Mentzelia involucrata*
pega pega   Sunflower Family

A short-stemmed flower, cream or lemon in color, this is one of numerous western wildflowers given the name blazingstar. Each flower emerges from a set of narrow-lobed bracts (leaves from which a flower arises) coming off the whitish stems. Although other stickleafs are found in flower along sandy washes in early summer or fall, this species adds its color to the desert between February and April and is spent by summer. In far western Arizona, this flower is mimicked in size and color by the Mohave ghostflower, *Mohavea confertifolia*, which is pollinated by many small bees that usually visit blazingstars.

Regions 1, 3

JUDY L. MIELKE

owl-clover

escobita   *Orthocarpus purpurascens*
Snapdragon Family

Owl-clover covers large areas in southern Arizona with a purplish glow for a few weeks each wet spring. This color is as much a function of the bracts beneath each flower as it is of the sepals and petals themselves. Its flowering spikes vary from hot purplish-pink, to velvety red-violet with yellow spots. Alone or in mixed stands, this annual inhabits plains, mesas, and slopes of deserts and grasslands between 1,500 and 4,500 feet. Early April is the best time to catch its color, but flowering may be just beginning then or may already be in its fourth week.

Regions 1, 2, 3, 4, 5, 6

WILLIS PETERSON

JERRY JACKA

STEVEN F. WARD

beardtongue

Penstemon parryi

Parry's penstemon, wind's flower   Snapdragon Family

A spray of these rose-magenta, funnel-shaped flowers can make any landscape especially memorable. They are even more attractive to bees and to Anna's and Costa's hummingbirds, which use them as one of their four or five major nectar sources in the spring. Flowering time is rather reliable, beginning the first week in March, peaking in late March, and lasting seven or eight weeks nearly every year. The flowers are on long stalks underlain by pale-green, arrow-shaped leaves. This species favors rocky bajada slopes, desert grasslands, and mountain canyons from 1,500 to 5,000 feet, but is often widely dispersed in its landscape, rather than clumped.

Regions 1, 2, 5, 6

SALLY WALKER

phlox

Phlox tenuifolia

desert phlox   Phlox Family

These perennial, slightly woody herbs form tufts or mats in open desert areas, with the flowers shooting up to heights of two to three feet. The white and lavender funnel-shaped flowers of the desert phlox are unique for this genus of plants and are probably pollinated by moths. They add sweet and musky scents to rocky slopes between 1,500 and 5,000 feet and are known only in Arizona. They bloom from mid-March to early May, peaking before April.

Regions 1, 2, 5, 6

PETER KRESAN

desert-chicory

Rafinesquia neomexicana

chicory, desert-dandelion   Sunflower Family

A white-flowered spring bloomer, this annual shares a few superficial features with the blue-flowered perennial chicory introduced from Europe and now widespread throughout much of America. Desert chicory is short but profusely branching and has deeply divided, arrow-shaped leaves. In wet winters, it becomes abundant upon plains, gentle bajadas, and mesas, from 200 feet in the desert, to its grassland edge above 3,000 feet. In some dry years, it may fail altogether even when other herbs flower. Its bloom may stretch from March to May.

Regions 1, 2, 3, 6

chia

desert chia, desert sage,
California chia

Salvia columbariae
Mint Family

Ranging from sky blue to indigo, the flowers of chias cluster like rings on terminal spikes above a highly-divided foliage. Popular with many bees, the heights of the plants vary from six inches to a foot and a half, depending upon moisture availability. The seeds of this annual contain a mucilage which gels up when they are placed in water, forming a sticky, inflated mass. This characteristic allows the seeds to stick to a site once rainfall has begun and perhaps provides germinating seeds with additional moisture. Chias bloom from sea level to 4,500 feet at the desert edge, from early March to July.

Regions 1, 2, 3, 4, 6

JIM HONCOOP

crownbeard

golden crownbeard,
cow pasture-daisy, butter-daisy,
hierba de la bruja, girasolillo

Verbesina encelioides
Sunflower Family

This long-season annual has triangular, toothed leaves that are grayish-green in color and fetid in odor. Crownbeard's ray flowers are a gorgeous yellow-orange, each three-notched at its end. The disk flowers are yellow, and are followed by flattened seeds covered with fine, gray-brown hairs, hence the name crownbeard. This plant colonizes disturbed roadsides and abandoned fields, from deserts clear into the pines. One set of these plants germinates early, to flower between March and July. Another, larger crop, germinates with the late summer rains and flowers from July through December.

Regions 1, 2, 3, 4, 5, 6

DESERT BOTANICAL GARDEN

Joshua-tree

yucca

Yucca brevifolia
Agave Family

The tall, shaggy-barked Joshua-tree produces gangly clusters of leaves that are relatively wide-ridged and have a coarse-toothed edge. Short flower stalks arise from the ends of its leafy branches. Forests of these plants are best known north and west of Wickenburg, on rocky plains and hillsides. The plant's short bloom time is seldom more than three weeks, but the simultaneous bloom of so many large, pale green flowers lights up the desert in March and April. The yucca moth gathers pollen, crams it on the stigma, and then lays its eggs in the developing fruit.

Regions 1, 3, 4

BOB CLEMENZ

Chapter 6
Late spring/summer wildflowers

Beginning in late April, daytime temperatures can exceed 95 degrees in the desert. Evaporation increases rapidly, and the superficial soil moisture disappears. The shallow-rooted ephemerals wither and die, leaving woody perennials and succulents to make up what little desert ground cover there is. Because the desert often receives no rain from late April through early July, it is surprising to some people that anything would bloom during this, the season of greatest water deficits.

Yet many deep-rooted trees and succulent plants in the Sonoran Desert do time their flowering so that ripened seeds will be ready to germinate as soon as the first summer rains come. This means, obviously, that their reproductive effort must occur *in advance* of the rains, rather than with them. Thus, even during the desert's harshest time, there is color, fragrance, and pollinator frenzy.

The prickly pear, cholla, saguaro, and night-blooming cereus are among the cacti that choose this season to bloom. Like the agaves and certain yuccas that bloom at this same time, they are much more efficient in their water use than are hybrid corns. Succulents are also 10 times more efficient water users than winter-flowering desert wildflowers.

The other major blooming components in late spring, the tree legumes, are deep-rooted members of the bean family. Some, like the paloverdes and ironwood, bloom very briefly at the start of this season. Others, like mesquite and desert senna, have one peak bloom in late spring and another in late summer, but they may develop a few flowers almost all the way through the warm season. Many ocotillo stands have a brief blooming period during this season.

A few herbaceous plants, such as desert-marigold, coyote gourd, sacred datura, and globemallow can be found in flower during this season, but they extend through other seasons as well. They have either deep or succulent roots and often grow in sandy washes where winter soil moisture lingers.

Regions 1, 2, and 3 have the most active flowering plants during this season, but Regions 4, 5, and 6 harbor a few species which peak in late spring. Above 4,500 feet, where soil moisture lingers, spring wildflowers may persist into early June.

To represent this season's bloomers, we have selected Desert Botanical Garden records of the following plants: saguaro, a columnar cactus; ironwood, a tree legume; and soaptree yucca, a succulent. Between localities, there may be as much as a three-week difference in the initiation of bloom of these plants.

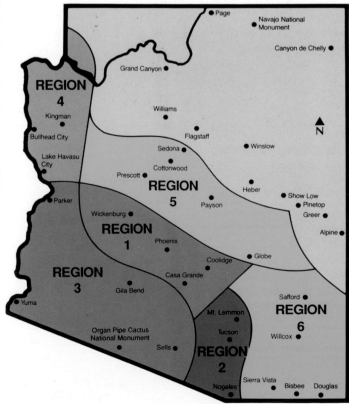

(LEFT) *The creamy white blossoms of the saguaro cactus are the Arizona state flower.*
KATHLEEN NORRIS COOK

Typical representative flowers of the late spring/summer bloom season

saguaro
woody perennial

Usually by the middle of May, saguaros in the Desert Botanical Garden are in full bloom. Late May to early June flowering is by far the most likely time to find saguaros topped with creamy white blossoms. (See page 70 for plant description.)

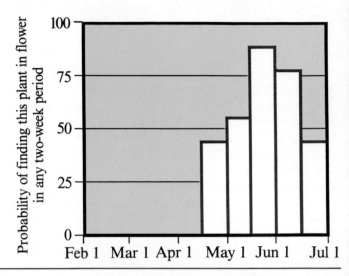

ironwood
woody perennial

Desert ironwood at the Desert Botanical Garden usually blooms from the middle to the end of May. The soft violet blossom is not showy but gives a dramatically different purple glow to the desert washes and slopes it inhabits. (See page 71 for plant description.)

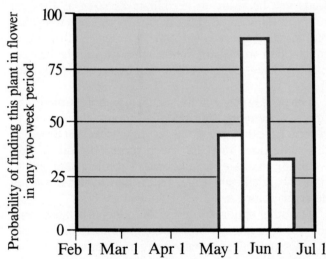

soaptree yucca
woody perennial

Records of soaptree yucca at the Desert Botanical Garden indicate that it usually blooms here in either late April or late May. Its bloom time sometimes even occurs as late as early June. (See page 74 for plant description.)

(FAR RIGHT) *A desert bouquet of beaver-tail cactus. The color is so intense that no printer's ink can equal its bright beauty.* JOSEF MUENCH

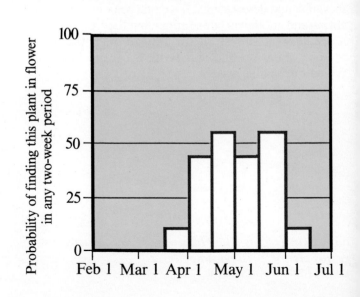

C. ALLAN MORGAN

C. ALLAN MORGAN

DAVID BURCKHALTER

saguaro
giant cactus, sahuaro

Carnegiea gigantea
Cactus Family

This unmistakable columnar cactus is the most familiar plant in the eastern Sonoran Desert. Bee researchers have shown that a saguaro produces an average of 300 trumpet-like flowers a season. Bats have lost out in the present day to pesticides and habitat loss, but bee pollinators have taken over and are now responsible for most of the fruit set in saguaro. Saguaros are largely restricted to rocky hillsides and coarse gravels on upper and lower bajada slopes. Three-fourths of the flowering is concentrated within six weeks, usually between the last of May and early July. However, a few blooms are spotted in early April, and infrequent stragglers open into the fall.

Regions   1, 2, 3, 4, 5, 6

blue paloverde
paloverde

Cercidium floridum
Bean Family

The blue paloverde is distinguished from foothills paloverde by larger, fewer leaflets, and by all-yellow flower petals with small red spots. This tree has a bluer cast than its foothills counterpart and is more restricted to plains and margins of dry washes. Both native paloverdes are now classified by some botanists as *Parkinsonia* species, because of hybridization with the introduced Mexican paloverde (*P. aculeata*). The blue paloverde has a very brief April-May bloom period most years, but aberrant individuals may make a few floral buds as early as February or as late as October. Most trees bloom synchronously over a 10-day to three-week period, peaking the last week in April.

Regions   1, 2, 3

foothills paloverde
littleleaf paloverde

Cercidium microphyllum
Bean Family

This small tree of upper bajadas has five to seven pairs of leaflets and bicolor (white and yellow) flowers on branches that end in a thorn-like point. Its bloom can be just as brief as that of the blue paloverde; a duration of less than 10 days is not uncommon. This period varies from one place to another, but ranges from early March through late May. The pale yellow flowers have white banner petals. During flowering, this tree's color can dominate entire desert mountain slopes.

Regions   1, 2, 3, 4, 6

WILLARD CLAY

sacred datura
toloache, jimsonweed, thorn apple

Datura meteloides
Nightshade Family

JERRY JACKA

A perennial datura, this widespread jimsonweed has fragrant white trumpet flowers with a purple or lavender flush in the throat. Its nodding, apple-shaped capsules are spiny and open irregularly to drop their buff seeds. Daturas, which are visited by both hawk-moths and honeybees, have a long but irregular blooming season. This species is a deep-rooted drought tolerator that inhabits sandy flats and arroyos. It puts on a new flush of flowers after any substantial rain between April and November. A diminuitive annual relative, *Datura discolor*, has a shorter blooming period and smaller range but is just as attractive. Both are poisonous.

Regions   1, 2, 4, 5, 6

ironwood
palo de hierro, desert ironwood

Olneya tesota
Bean Family

C. ALLAN MORGAN

CHARLES BUSBY

The greyish-green leaflets, combined with blossoms varying from deep violet to white, give a pale purplish cast to the washes and slopes where ironwood lives. Ironwoods obtain heights of 30 feet and produce one of the heaviest woods in the world. This fine hardwood is used for radiantly polished woodcarvings of desert and sea animals by the Seri Indians. At the northern edge of their range, ironwoods freeze often and become more sparse. They seldom extend above 2,500 feet. Bee pollinated, they have an extremely compact flowering season ranging from three weeks to as little as six days. This season begins in mid-March on the Mexican border but is initiated later, around mid-May, farther north or at higher elevations.

Regions   1, 2, 3

buckhorn cholla
Opuntia acanthocarpa
Cactus Family

JOHN CACHERIS

Heavily armed by long spines, each with a sheath-like covering, this forbidding cholla offers dazzling red, yellow, orange, or variegated blossoms. They spiral inward when touched, which results in bees getting doused with more pollen. It has joints six to 12 inches long and spine-bearing tubercles that are two to three times as long as they are wide. The open, branching habit makes it easy to pick the buds, which are boiled or steamed in pits. The cooked buds have the flavor of asparagus or artichoke. Found from 500 to 3,500 feet, sometimes hybridizing with other chollas, buckhorns are plants of open valleys, bajadas, and arid canyons. Bloom time runs from mid-April to late May.

Regions   1, 2, 3, 4

DESERT BOTANICAL GARDEN

pancake-pear   *Opuntia chlorotica*
silver-dollar cactus, nopal, tuna   Cactus Family

The flat, circular joints of these cacti are yellow-green in cast, with a mat of fierce bristles and yellow translucent spines to greet the uninitiated. The pads produce yellow flowers and fleshy fruits along their margins. Pollinated by solitary bees, these prickly-pears, like most others, have numerous bright stamens which swirl into a tight coil around the stigmas when touched by insects (or by fingers). Found from 2,000 feet in the desert up to 6,000 feet in the pinyon-juniper woodlands, this cactus-pear is always in bloom by mid-May, but some years begins as much as six weeks earlier.

Regions   1, 2, 3, 4, 5, 6

R. E. SCHNITKEY

Engelmann's prickly-pear
nopal, tuna   *Opuntia engelmannii*
Cactus Family

The most common prickly-pear in Arizona deserts, this species is also one of the most variable. Its branches of pads may be erect, ascending or sprawling; the joints may be six to 10 inches long; the spines may be flattened, straight or curved, brown or white. Flower color may vary from yellow to pink to red but they are all bee pollinated. This species extends from 1,000-foot deserts up to ponderosa pine and Douglas fir forests at 7,500 feet. Its desert bloom time comes up around mid-April and wanes in early or mid-June. Relatively reliable in flowering duration, the same individual plants will flower for about the same five or six weeks every year.

Regions   1, 2, 3, 4, 5, 6

DESERT BOTANICAL GARDEN

purple prickly-pear   *Opuntia violacea*
Cactus Family

With purple-tinged pads and opaque, red-brown spines, this prickly-pear is easy to distinguish from other species. Against these red-violet hues, brilliant yellow flowers are enchanting. One variety has red centers to its golden flowers. As far as we know, this is the only Arizona prickly-pear that provided floral buds rather than fruit as its contribution to Indian nutrition. Purple prickly-pears are found from 2,000 to 5,000 feet in Sonoran and Chihuahuan deserts and the grassland-woodland bridge between them. Flowers begin to open in early April, continuing through late May. Bees must crawl beneath many pollen-laden stamens to reach the nectar.

Regions   2, 5, 6

mesquite
velvet mesquite

Prosopis velutina
Bean Family

Perhaps the most common Sonoran Desert tree, mesquite is known nationwide for fine furniture, firewood, and for the mild-flavored honey produced by the bees that favor its flowers. An average-sized tree produces 12 million flowers per season. This species has slightly curved, often speckled pods with sticky beads of sugary sap on them and 12 to 20 leaflets nearly touching one another on each of two to four compound leaves. Roots run deep, often reaching underground water. The pods of this nitrogen-producing tree were the single most important food of Sonoran Desert Indians. The velvet mesquite bloom commences in late April, wanes by early June, then reinitiates in early August. Mesquites are somewhat sensitive to long freezes.

Regions   1, 2, 3, 4, 5, 6

C. ALLAN MORGAN

elderberry
desert elderberry,
tapiro, Mexican elder

Sambucus mexicana
Honeysuckle Family

As a large tree with opposite, divided leaves and flat-topped clusters of white blossoms, this Arizona wildflower is quite distinct from all others so far mentioned. This is the only elder in the state which reaches beyond the mountains. It has many different pollinators, making it a generalist. Both the flowers and fruits have been used as folk medicines, and when cooked, the fruits are edible or suitable for wine-making. The Mexican elder flowers from March into early July. It frequently grows next to streams and irrigation ditches.

Regions   1, 2, 3, 4, 5, 6

DESERT BOTANICAL GARDEN

trumpetflower
palo del arco, tronadora,
miñones, esperanza, yellow trumpet

Tecoma stans
Bignonia Family

A multi-stemmed shrub with shiny leaves divided into five arrow-shaped leaflets, trumpetflower showers its landscape with golden flowers much of the year. In frost-free areas, it grows into a large tree. Trumpetflower roots continue to be used medicinally in Mexico. It prefers dry, rocky, or gravelly slopes below 5,500 feet in deserts or in grasslands and woodland canyons which drain into deserts. The flower has a sensitive stigma, which slams shut like a clam when touched. Trumpetflowers begin to bloom and attract large bees in late April, flowering sporadically into November or December.

Regions   2, 6

JAMES METCALF

C. ALLAN MORGAN

soaptree yucca
Spanish bayonet,
palmilla, soapweed yucca

Yucca elata
Agave Family

A tall, multi-branched yucca, soaptree has narrow leaves with thread-like margins and dry capsules which split open and wind-disperse their seeds. The creamy white flowers are produced on long stalks. They are pollinated by specialized moths. Soaptree roots historically were used as a shampoo. The weaving of the leaves into Indian baskets persists until this day. A grassland species, soaptree extends farther into the Chihuahuan Desert than into the Sonoran Desert. It blooms in June and July in southern Arizona and in early April and May at lower elevations in central Arizona.

Regions   1, 2, 5, 6

(LEFT) *The desert is home to many creatures, including this pair of young hawks whose nest of sticks is secure in the spiny arms of a giant saguaro.* WESLEY HOLDEN

(BELOW) *The bloom of the ironwood is relatively brief. But when it happens all at once, as in this picture, it turns this desert hillside of Saguaro National Monument an incredible smoky-lavender color.* PETER KRESAN

Chapter 7
Late summer/fall wildflowers

During the long, hot days of midsummer, flat-bellied clouds build up over the Sonoran and Chihuahuan deserts. Moisture-laden air currents from the Gulf of Mexico sweep up into Arizona, raising the humidity and creating a scatter of thunderclouds that storm over the land. Rains are preceded by dust storms and violent lightning.

When the hard, brief rains hit, they churn up the seeds of many annuals and perennials that have been hidden in the soil. In the hot, drenched surface layers of desert soil, another set of desert plants begins to germinate. It emerges within days, if not hours, of the first summer thunderstorms heavy enough to cause flash floods in desert washes.

Perhaps Arizona caltrop, amaranth, devil's claw, limberbush, morning glory, and desert zinnia best represent this warm, long-day, highly humid season. These annuals, which flower in late summer, have their affinities in the tropics. Many are rank, lanky, leafy annuals with large leaves or highly divided leaflets that track the sun when water is available and close up their surfaces when it is not. They may use a form of photosynthesis that works at high temperatures and high water availability to produce green matter at an extremely rapid, efficient rate.

Amidst all this greenery, flowers are often obscured during this season. Most of these plants need shortening days to trigger their internal clocks to initiate floral buds, but a few species will blossom now. The appropriate temperature with day length helps to make a significant showing.

A few cacti peak in their flowering cycle during this warm, wet season: pincushion cactus, jumping cholla, and certain barrel cacti. Up at the desert edge in mountain canyons, a number of shrubs flower in profusion during late summer, although they have been modestly active for months: wild cotton, desert-willow, hibiscus, trumpetflower, and desert senna.

Regions 2, 5, and 6 gain the most benefit from the thunderstorm fury of late summer. Several species of morning glory, devil's claw, buffalo gourd, honeysuckle, sunflower, nightshade, tomatillo, caltrop, bee balm, and wild bean sprinkle color into the grasslands and savannas.

The summer wildflowers seldom appear as one large chunk of color. They are scattered, and become concentrated only along the edges of roads and washes. When they begin to fade, desert broom floods these landscapes with its cottony seeds, and wild golden gourds can be seen hanging like ornaments from trees where their vines had crawled earlier. Groundsels, paperflowers, brickellbushes, telegraph plants, burroweeds, and other composites also help finish off the season. To represent the late summer/early fall bloomers, we have selected the following three species: Arizona caltrop, an annual; barrel cactus; and wild cotton, a shrub.

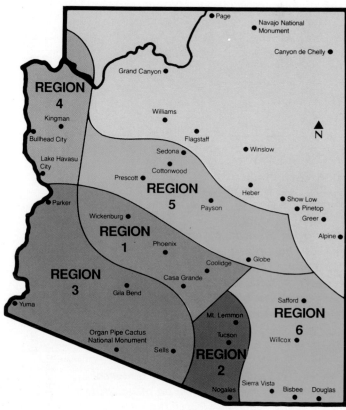

(LEFT) *Small, and often solitary, the pincushion cactus finds protection under large cholla.* JERRY JACKA

Typical representative flowers of the late summer/fall bloom season

Arizona caltrop
herbaceous annual

The most likely time to find Arizona caltrop flowering in the Desert Botanical Garden is from the middle of August through September. Some years it has begun flowering in July. (See page 83 for plant description.)

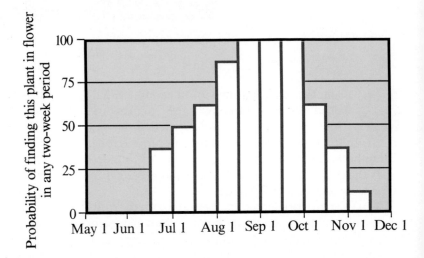

barrel cactus
perennial cactus

The same cactus does not bloom twice, however different populations of barrel cactus may have different bloom times. Each population will usually bloom with late summer rains, or in some years as early as mid-June. (See page 81 for plant description.)

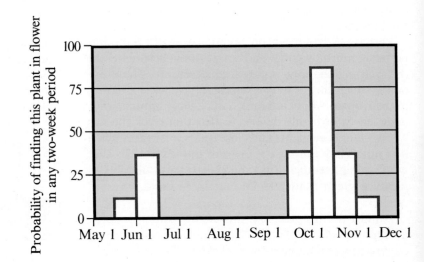

wild cotton
woody perennial

Mid-August through September is when the flowering time for wild cotton usually occurs in the Desert Botanical Garden. However, in some years the plant has been recorded as blooming as early as late May on records kept between 1978 and 1986. (See page 82 for plant description.)

(FAR RIGHT) *Visitors to the Desert Botanical Garden enjoy a peaceful stroll on meandering garden paths.* J. PETER MORTIMER

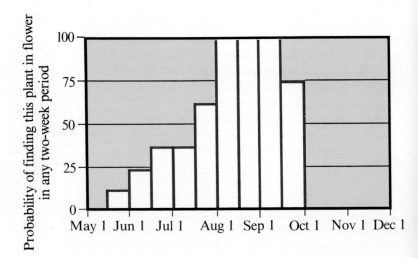

JUDY L. MIELKE

TIM WALKER

WESLEY HOLDEN

century plant
Parry's agave, mescal

Agave parryi
Agave Family

This century plant takes more than 25 years to send up its 12- to 18-foot flowering stalk, then blooms and dies. Its wide, spatula-shaped leaves make compact heads with a pale gray-green cast to them. The yellow stamens protruding from their blooms add a brightness to the subdued grasslands and woodlands. Although other large panicled agaves are bat-pollinated, this species extends beyond the range of nectar-feeding bats and may be pollinated by bees and hummingbirds. The species ranges to 9,000 feet in elevation, suggesting considerable freeze-tolerance. It blooms mid-June to mid-August, with entire clonal populations flowering synchronously.

Regions   5, 6

saiya
temaqui

Amoreuxia palmatifida
Cochlospermum Family

These herbs rise, after the summer rains, from a perennial tuberous rootstock and persist above ground less than three months. The orange flowers cluster above or between the hand-shaped leaves. Lacking nectar but still showy, the five petals have brown spots at their bases and numerous stamens, which drop their pollen when bees buzz or vibrate their wings nearby. Every part of the plant is edible, from parsnip-like roots to seed capsules which are used as a coffee substitute. Found in hills and canyons from 3,000 to 6,000 feet, it flowers July through September.

Regions   2, 3, 6

desert-willow
desert-catalpa, jano, mimbre

Chilopsis linearis
Bignonia Family

Although this tree has leaves like a willow, its flowers smell more like violets and look more like orchids. White, lavender, and yellow background colors are sprinkled with dots, blotches, and streaks representing the whole spectrum. These large blooms are perched on their branches in a manner which makes them perfect aids to bee and hummingbird-watching. Although they frequent desert washes, desert-willows become abundant at higher elevations in grasslands and oak-juniper woodlands. Blooming seldom begins by mid-May and may end by early summer or add new flushes until early October.

Regions   1, 5, 6

coyote gourd

fingerleaf gourd, chichicayota,
calabacilla

Cucurbita digitata
Gourd Family

BILL CRAMER

This warm season vine emerges prior to summer rains from huge water-storing taproots. One form, with finger-like lobed leaves, hybridizes with a broader leaved form found from the Yuma area westward into California. The large, pale yellow flowers open before dawn and are pollinated by at least two kinds of squash and gourd bees. Both male and female bees visit these flowers, the males patroling for mates and sleeping in the flowers while they are closed for the night. Blooms are followed by large round gourds, green-striped or yellow. The roots and gourds have been used medicinally for millennia. Flowering time occurs as early as May and as late as October.

Regions 1, 2, 3, 4, 5, 6

coral bean

chilicote, colorin, patol

Erythrina flabelliformis
Bean Family

JACK W. DYKINGA

Coral beans are peculiar in the Arizona flora in that they begin to bloom before their shiny, aspen-like leaves appear on their short, woody branches. These shrubs begin to bloom in early June, peaking two to three weeks later and terminating completely by late July. Leaves and red-beaned pods begin to develop as the summer rains come, but the leaves seldom persist more than two months. Black-chinned hummingbirds are among the pollinators of the sucrose-rich tubular flowers. The blood-red seeds are rich in toxic alkaloids, which, along with ants, partially protect the seeds from consumption by moth larvae. Coral bean is found from 3,000 to 5,000 feet in desert grasslands and oak-pine woodlands.

Regions 2, 3, 6

barrel cactus

visnaga, biznaga,
fishhook barrel, Wislizenus' barrel

Ferocactus wislizenii
Cactus Family

G. C. KELLEY

This rotund cactus has dark, hooked, flattened spines with bristles between them. It is one of the taller species of barrels. While it can store much liquid in its succulent flesh, it responds rapidly after a rain, immediately putting out new root hairs to capture new soil moisture. It blooms with the summer rains, beginning by mid-June and peaking in late August. The cup-shaped flowers range from red to yellow but most are orange. They have little fragrance, but attract many bees. The flowers open mid-morning. Delicious yellow fruit are produced.

Regions 1, 2, 5, 6

GARY P. NABHAN

GEORGE H. HUEY

C. ALLAN MORGAN

wild cotton
algodoncillo, desert cotton

Gossypium thurberi
Mallow Family

Since this relative of cultivated cotton also serves as an alternate host for the boll weevil, it is often removed from nearby fields. Yet its lovely five-lobed leaves and pale pink to whitish flowers justify its protection for ornamental value where it grows apart from cotton fields. Many bees and wasps are attracted to its floral nectar, but it also exudes nectar on the leaves and flower stalks, thereby attracting other insects. Preferring rocky slopes and washes, it ranges from high desert at 2,500 feet to oak and juniper woodland at 5,000 feet. It is known to flower as early as April, but most blooming and fruiting is from August to October.

Regions   1, 2, 3, 6

snowy sunflower
girasol,
silver-leaved sunflower

Helianthus niveus
Sunflower Family

This true sunflower grows only in sand dunes, swales, and roadsides adjacent to them. It can develop into four-foot-tall, many-branched plants, or produce six-inch miniatures and mature at either size. Snowy sunflowers can germinate when there are rains in winter or late summer, flowering either from February through May or in November. Enduring tremendous heat, this desert herb is more efficient at using water than giant sunflowers or any of their relatives. Plant breeders are using it to increase the drought and temperature tolerance, oil seed quality, and insect resistance of cultivated sunflowers. It is pollinated by a wide variety of bees, and other insects.

Region   3

desert rosemallow
Coulter's hibiscus, pelotazo

Hibiscus coulteri
Mallow Family

The spectacular mix of colors in the large flowers of this straggling shrub can stop any hiker dead in his tracks. Each delicate yellow petal has a blood-red basal spot amidst other markings. Many kinds of bees find this flower attractive, although certain species specialize in the mallow group. Its slender wood branches reach four feet in height, but it never develops a large canopy. Desert rosemallow blooms sporadically throughout the year, even during late fall-early winter droughts. It inhabits steep canyons from 1,500 to 4,500 feet in desert and desert grassland.

Regions   1, 2, 3, 5, 6

Arizona caltrop   *Kallstroemia grandiflora*
orange caltrop, summer-poppy,   Caltrop Family
Arizona-poppy

This summer bloomer has five orange petals, each with pale red veins. Numerous flowers bloom simultaneously on this ground-creeping herb with divided leaves. It germinates with the first summer rains, then sprawls out in wet pockets along roads and washes to cover whole patches with luxuriant growth. Quick to bloom once established, its desert and grassland flowering season may extend from early July through October, but most years it lasts only from late July to mid-September. Bees are active pollinating this plant not because they like the pollen; they groom it off when it sticks to them.

Regions   1, 2, 3, 6

JOSEF MUENCH

pincushion cactus   *Mammilaria microcarpa*
Cactus Family
fishhook cactus, cabeza del viejo

The small cylindric stems of this cactus produce pink to lavender blooms with white margins, pollinated by bees. The scarlet, berry-like fruit which follow are as colorful as the flowers themselves. Although their hooked central spines offer some protection against browsers, this species tends to grow beneath chain fruit cholla, thereby avoiding animal feeding and trampling. Stems are often solitary, or sometimes in small clusters. It is common on both heavy and light soils, on rocky slopes and plains of deserts, and grasslands below 4,500 feet. Its flowering range is from mid-April through early September, but it appears to be triggered by summer rains, putting out new flowers five to seven days after each drenching.

Regions   1, 2, 3, 4, 5

C. ALLAN MORGAN

jumping cholla   *Opuntia fulgida*
chain fruit cholla, choya   Cactus Family

A tall, thick-stemmed cholla, this species lacks the knob-like tubercles on its fruit that are present in most other chollas. Its long, spiny, inch-thick joints are readily detachable with the slightest pressure, hence their apparent "jumping" nature. The bee-pollinated flowers are small but a deep pink. Flowers and fruit tend to hang upside-down from pendant branches of this tree-like cactus. Flowering runs from the dryness of mid-May, through the mid-summer rains, into late August. The peak is between late July and early August. Below 4,000 feet, jumping chollas cover a wide range of habitats but form impenetrable stands along the U.S.-Mexican border west of Organ Pipe Cactus National Monument.

Regions   1, 2, 3

LON McADAM

ANDREA MARIE VINCENT

C. ALLAN MORGAN

night-blooming cereus
reina-de-la-noche,
Arizona queen-of-the-night

Peniocereus greggii
[Cereus greggii]
Cactus Family

The waxy-white, 6-inch-long flowers of this tuberous cactus produce a fragrance that can be smelled 100 feet away. Beginning just after dusk, during June and July, the buds unfold quickly, in spasms, to attract moths and other insects throughout the night and until dawn. From one to 25 flowers can open on a plant the same night, but rarely does an individual's total floral production exceed fifty blossoms per year. The flowers are produced on slender, lead-colored stems which often camouflage themselves under creosotebushes.

Regions   1, 2, 3, 6

GARY P. NABHAN

CHARLES MANN

devil's claw
unicorn plant, guernito,
uña de gato, torito

Proboscidea parviflora
Unicorn Plant Family

The reddish purple, pink, and yellow-striped tubular flower of this species attracts large bees, which trigger the sensitive stigma to "slam shut" a few seconds after pollen is deposited. After the fruit matures from its okra-like green stage, it sloughs off its skin, splits down the middle, and two horn-like projections curl back. Southwestern Indians have domesticated a variety with unusually long horns, whose fibers are woven into baskets for design elements. On wet roadsides or in washes, annual devil's claw will sometimes begin to flower in mid-May, but the bulk of the plants germinate with summer rains and kick into flowering within three weeks. This second flush in mid-July may last through mid-October. The range of annual devil's claw has been greatly extended by Indian trade and livestock so that it now stretches from 500 to 5,000 feet elevation.

Regions   1, 2, 3, 4, 5, 6

desert senna
senna, dais, rosemaria

Senna covesii
[Cassia covesii]
Bean Family

The desert senna is an herb with rusty-yellow flowers rising from woody perennial woodstocks. It inhabits dry, rocky slopes and mesas, adding golden hues to early summer and late fall when few other shrubs bloom. Nearly every year, it can be found flowering somewhere between mid-April and mid-November; sometimes a few flowers open in March or even earlier. The nectarless flowers are "buzz-pollinated" by bees which use their flight muscles to vibrate the pollen free from the anthers.

Regions   1, 2, 3, 4, 5, 6

JERRY JACKA

desert zinnia
wild zinnia, zacate pastor

Zinnia grandiflora
Sunflower Family

A low perennial with clusters of bright yellow flowers, the desert zinnia is a warm season bloomer. Its opposite, three-ribbed leaves underlay a profusion of blooms from May through October. Its pollen attracts a considerable number of honeybees from August onward. This species is a grassland and oak woodland dweller but does well when cultivated in the desert. Preferring elevations between 4,000 and 6,500 feet, it grows on semi-arid slopes, mesas, and plains.

Regions   2, 6

CHARLES MANN

(BELOW) *Summer brings a brilliant crown of blossoms to this barrel cactus in the Ajo Mountain foothills in Organ Pipe Cactus National Monument.* JOSEF MUENCH

Planting and nurturing wildflowers can be a tangible pleasure. We can watch a dozen kinds of wildflowers sprout, bloom, and seed within a few feet of where we eat and sleep, tying our lives into the soil and seasonal changes. SCOTT SONES

Chapter 8
Return of the native:
desert wildflowers in and around the home

If you've ever wandered through a yard devoted to native annual wildflowers, trees, and shrubs, you may have walked away with a sense of harmony. The setting is in synch with the underlying nature of its surroundings. That compatability is often forgotten in a lavish but contrived urban landscape, a Mediterranean-flavored suburban subdivision, or even in a single-crop farm.

The native setting has the color, the fragrance, and sometimes the wildlife that could be found on the same site one hundred or even one thousand years ago. Such native plantings soon become less of another human contrivance and more a blending with natural settings much larger and older than all of us.

Of the many reasons for sowing natives, this one seems foremost: having native wildflowers around our home reminds us of the uniqueness of our local setting and makes that distinctive sense of place part of our lives.

Certain natives require less water and less pampering than commonplace exotics. Since outdoor uses account for more than half the water consumed in southwestern cities, such excesses should hardly be tolerated in a region where this resource is already limited.

Other reasons are more abstract. When we show appreciation for the desert flora, rather than overlooking it, we are voting to recognize its intrinsic worth, thereby encouraging its conservation in its wild setting. As desert ecologist Paul Sears observed many decades ago, "People seldom try to save anything that they haven't yet learned to love." By becoming familiar with desert plants in general, we may be more likely to protect particular rare ones threatened by our own activities.

Planting and nurturing wildflowers can be a tangible pleasure. We can watch a dozen kinds of wildflowers sprout, bloom, and seed within a few feet of where we eat and sleep, tying our lives into the soil and seasonal changes. Fortunately, within the last decade, the kinds of desert wildflowers available through nurseries and seed catalogs has increased tremendously. Much new information has emerged regarding their care and feeding. It is easier than ever before to return our gardens to the natives, or to heighten the beauty of a yard where natural vegetation already exists. Guidelines follow for getting started with native landscaping. Other information in this book will provide additional suggestions for help once you are hooked.

DON B. STEVENSON

A. Knowing your conditions

Perhaps the most important preparation for a wildflower garden is in matching the site you choose with the wildflower array suited to your area. Get to know the various parts of your yard from a wildflower's point of view.

What is growing on it now and how well do these plants fare over different patches of soil? What do old-timers or historic descriptions of your town say about what grew on it in decades past? How well does the ground hold moisture? Is the soil surface slick after a rain but quick to dry again? Is there much runoff? Does caliche hardpan limit deep root growth? How much of the day does sunshine reach one spot versus another? Does frost hit one side of the yard harder than the rest?

Once you have discovered the "micro-environments" in your own yard, you will begin to notice wild plants which comfortably grow under natural conditions not unlike the topographic niches in your yard. Using this book and others, keep a checklist of natives potentially suited to your own setting. Walk around the vacant lots, desert washes, or open lands nearest to your home. What wild species do well there that can be added to your own site? (If neighbors want to share seed from wildflowers volunteering on their own land, this may be just as good a source as any.)

You may not necessarily want all future plantings to adapt to your lot *as it is*. Maybe it is worth terracing an eroding hillside, or digging out old building rubble and refuse from a trash pit, building up fertile organic matter in their stead. For the calcium-carbonate hardpan known as caliche, you may even want to rent a jackhammer or soil auger to crack through subsurface layers in order to give tree and shrub roots more room to grow. Removing bermudagrass may also be necessary, and can be done either by physical or chemical means.

B. Learning what you want: function and formality

How do you use your yard? Yards seldom function solely for aesthetic pleasure. Both native and non-native plantings may get in the way of certain activities if inappropriately located or improperly maintained. Newcomers to the desert naively dismiss native plants as being incompatible with family life around the yard, even though exotics (introduced ornamentals) may be just as incongruous in some settings. It's true that you don't want to plant prickly pear in the midst of where the kids play football, but neither will you want to plant the spiny, non-native pyracantha. On the other hand, native grama grasses and touch football may make a great team.

If you don't need to reserve space for an occasional ballgame or have no predilection for formal garden plantings but love birds and flowers, you may opt for a more natural planting design.

In the northern U.S., these have been collectively termed "wildflower meadows," whether they are tall grass prairies replacing trimmed lawns or bog gardens modeled after the flora of primeval wetlands. The goal is a mix of plants blooming at different times of the year that replicates a natural habitat of the region in terms of life forms, plant spacing, and diversity.

This naturalistic approach is a reasonable way to treat areas that are still relatively pristine and rich in indigenous plants and animals. Since a diminishing number of places have a "desert feel" to them, why remake them as well? A similar strategy, called "restoration ecology," enhances lots which still have remnants of the preexisting vegetation, complementing these with formerly-present species. You simply weave colorful species back into the slightly-worn

fabric of nature's quiltwork. For instance, reintroduced wildflowers can be sown beneath the canopy of taller mesquites and clusters of cacti or subshrubs planted nearby.

Yet it is a fallacy to consider native plants as being suitable *only* in such wild settings. As Texan Jill Nokes has observed in her horticultural guide, *How to Grow Native Plants of Texas and the Southwest:* "Because few people want the thicket or wild look in their yards, they believe they must forgo native plants altogether…But native plants can and should be used as single specimen plants, formal hedges, or ground covers…or as perennials mixed with other exotic bedding plants."

Nokes' observation is illustrated in many formal landscape designs scattered throughout the Southwest and northern Mexico. Clusters of yellow trumpet flower shrubs are pruned into beautiful courtyard carousels in Baja California. Large agaves or barrel cacti, set into volcanic rocks and cinders, and angled in different directions, make elegant, sculpture-like focal points in plazas. Hedgerows of Joshua-tree, prickly-pear, or sotol can add much more texture to a desert landscape than can a row of non-native oleander or privet.

Even when a family opts for a formal patio or porch, planters and beds full of native annual wildflowers can offer life that overused pink petunias can't come close to. Overall, it is best to think of how you want a flower bed, windbreak, sound barrier, or sidewalk margin to function, then scan the wide range of natives and select one for the job. In most cases, there is a low-water-using, high-heat-tolerant, indigenous plant that can fit the bill once you decide what color, size, and shape you want.

C. Obtaining the plants you want: seeds and nursery stock

Some people will keep a plant shape or color in their memory for years before they realize it fits a need in their yard. At that point, the problem is how to make that image match a name in a seed catalog or a sprout in a local nursery.

The earlier chapters in this book describe qualities of 63 selected species of the hundreds of natives that grow in Arizona. For an updated list of wildflower seed and plant sources for these natives, contact the Desert Botanical Garden Library in Phoenix or the National Wildflower Research Center in Austin, Texas. Other sources exist, and local nurseries should be consulted. Also helpful will be the Planting Guidelines (see pages 104-107) for planting Arizona desert wildflowers.

Once you find a source of seed or nursery stock, it is best to read the detailed planting descriptions provided by the supplier. Sometimes, a California strain of a "native wildflower" sold in Arizona may bloom at a somewhat different time than the usual blooming period of an Arizona strain. When planting information from a seed catalog does not correspond with this book's Planting Guidelines, don't condemn anyone. Experiment! Try them both ways, and record your results to share with other native plant enthusiasts.

PHOTOGRAPHS BY DON B. STEVENSON

If we include showy native trees and shrubs in our definition of wildflowers, additional factors must be considered. Sometimes there is a choice of starting a flowering tree or shrub from either seed or from nursery stock. If a local nursery has large-sized plants in five- or 15-gallon containers, and you are in a hurry for "appearance," the nursery material offers immediate advantages.

However, you may not be able to find all the varieties native to your locality, since many perennials in nurseries are started from cuttings. These cuttings are not only genetically uniform, they are identical to the mother plant from which they were pruned.

PHOTOGRAPHS BY DON B. STEVENSON

Other things being equal, cuttings have a head start in growth over a seed planted at the same time. However, horticulturalists find that after a short while, seedlings that are progeny of cross-pollinations will be vigorous enough to exceed the growth rate of a cutting. For some species, cuttings are used because seed germination is difficult or seedling mortality high. For others, such as many hardwood trees, cuttings don't "take" under normal conditions, so breeders rely exclusively on seeds.

When you desire a species for which no seed or nursery source is available, you may have to rely on your own seed-collecting abilities. *With permission of the land manager or owner*, you can probably collect seeds of all but the rarest species without depleting that population's own seed reservoir for future reproduction.

Once you locate a population of a plant that you desire, determine how much filled, maturing seed is available before you collect any at all. Crack open shells, feel pods for fullness, or look for dark, rich hues in berries to determine if they are ripe. After assessing how many maturing seeds are available in a stand, decide to take no more than one to five percent of them, leaving nature some to recover with. Sort out insect-infested or diseased seed from healthy-looking ones. Collect them in paper or cloth rather than plastic bags, label their place of origin and species name or description in indelible ink on the bags, and carry them home for cleaning.

In most cases (except for some berries and acorns), you will want to clean and dry these seeds or fruits immediately. Dry them on screens in the open sun, turning them frequently, rather than in an oven or solar oven-dryer. Another alternative is to place the seeds with an equal or greater volume of silica gel dessicant in a sealed chamber — an empty terrarium or cardboard box — for two days. Once dry, most desert seeds can be stored at low temperatures for years without loss of viability, preferably in a low-humidity refrigerator or freezer. Keep their original identification label with them, whether you store them in paper coin envelopes, glass jars, or humidity-free "freezer bags." With seeds properly screened, cleaned, stored, and labeled, you can add them to your landscape at your leisure.

Although most seeds from desert environments are not normally exposed to freezing temperatures very long in nature, freezing or refrigerating will not hurt them. Unlike northern and high-elevation seeds, no Sonoran Desert species seeds that we know of need to be "cold-stratified" — placed outdoors in cold soil or in sand and moss (sphagnum) layers in a refrigerator — to enable germination.

In fact, just the opposite may be needed. Certain spring-blooming Arizona wildflowers do tend to germinate in higher percentages if they first experience a season on or slightly beneath the hot desert soil surface. There, where temperatures reach from 120° to 170°, they apparently reach physiological maturity without heat damage and readily germinate when exposed to moisture after temperatures cool again.

D. Preparing the seedbed: soil, water, and sowing

A loose, crumbly, or "friable" soil is best for encouraging seed germination. In that way, seeds can be in contact with moistened soil particles needed to encourage germination and sustain seedling turgor (swelling caused by water pressure). Seedlings can sink their roots, and their shoots can break through to light. Soil that is uncompacted, or perhaps previously worked, is preferable.

In any case, the ground should be rototilled or shovel-dug to a depth of eight or nine inches. This should be done one or two months in advance of planting. If noxious weeds have occurred on the site in years past, we recommend watering the area, letting these weeds emerge, then removing them once or twice before planting wildflowers. In this way, you reduce the exotics' seed reserve in the soil, and learn what their seedlings look like before you open the same ground to your wildflower friends.

In opening the ground, you risk the danger of competition from weeds. You may need to reduce this competition. For small areas, manual removal of weeds is sufficient, but professionals often rely on a mix of mechanical removal and herbicides — either pre-emergent chemicals targeted at certain weedy species, or post-emergent general herbicides. If you are considering the use of herbicides, keep in mind that certain chemicals also kill desirable plants, so take steps to minimize impact to human health and to wildlife.

Sprinkling the bed once or twice before wildflower planting also gives you a chance to test the soil's water-holding and infiltration capacity. If it is a heavy clay soil, till in a one-inch layer of sand. Adding two or three inches of organic matter, in the form of compost or peat, may also be helpful in increasing soil water holding capacity but may be too costly and generally inappropriate for larger areas.

Although many competent horticulturalists recommend manure and nitrogen fertilizers for wildflower beds, recent research suggests that such treatment may be unnecessary or even harmful for some wildflowers. For one thing, some nitrogen fertilizers may encourage prolonged vegetative growth at the expense of early flowering. Ammonium phosphate or other high phosphorus fertilizers may actually encourage flowering more than those with high nitrogen to phosphorus ratios.

Second, inorganic nitrogen inhibits several kinds of beneficial microorganisms which coexist with plant roots and thereby reduces the services which these symbiotic bacteria and fungi provide for plant growth and survival.

Finally, one attraction of desert natives is that they have adapted to nutrient-poor soils with high alkalinity and thus do not need the pampering and expensive maintenance required by nutrient-guzzling exotics.

Just before planting, level the soil, and rake it smooth. Then measure the total area of the bed you have prepared and calculate the amount of seeds (in weight) you should sow for an area that size by multiplying or dividing the estimates recommended on the seed package or in other instructions, or in the Planting Guidelines on pages 104-107. If you plan to mix different kinds of wildflower seeds in the same bed, reduce the density of each to a degree, or else there may be too much competition among them.

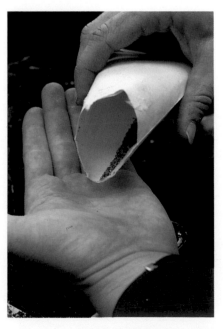

Although there are premixed regional wildflower seed packets, these vary greatly in quality from one seed company to another. Many mixes are carefully chosen to provide a varied palette of color for your yard, and their seed packets have introduced many to the joys of wildflower cultivation. Although we admire the efforts that seedsmen have made to introduce such mixes to the public, gardeners trying them for the first time should not expect too much.

Premixed wildflower seed packets marketed for "regional" designations are so general that you seldom obtain more than one or two species that are actually native to your locality. On packets of many southwestern regional wildflower mixes, several climatic regions within Arizona, plus others in adjacent states, are lumped together as "the Southwest." Natives from some of Arizona's regions are not represented in the mixes at all. These mixes also tend to lump together plants which flower in different seasons; planting them at the same time is not always helpful.

Another consideration is that certain of these regional mixes of "natives" have a high percentage of introduced alien weeds in them. It is safer and more

creative, though a little more costly, to mix your own array of wildflowers for a particular bed and a particular season.

Not all the seeds you plant the first year will come up in that season, so plan to use that bed for the same mix over several years' time. Different species of wildflowers have different sets of germination inhibitors, some of them chemical or physical coatings that need to be abraded away. To obtain some germination the first year, most species can be slightly abraded by shaking the quantity of seed ready for planting in a glass jar with rough-edged gravel and wet sand. Four or five minutes of this abrasion or scarification simulates what a flash flood does to soften seeds for germination.

Some propagation specialists recommend an overnight soaking of the hard-to-germinate seeds in water, or a shorter soak in hot water bath or in various solutions. Some dormancy-breaking solutions may be made from products regularly found in your kitchen, such as one made with a single part household bleach and 10 parts warm water; others use sulfuric acid or more obscure (and potentially hazardous) laboratory chemicals.

Additionally, germination of some species may be inhibited by light conditions. By placing seeds prior to planting under alternating light and dark, you can encourage species such as saguaro cactus to germinate more readily.

Broadcast this seed-gravel-sand mix evenly over your newly raked bed. Tamp down the seeds, putting them into contact with soil particles, by gently walking over the bed or pressing them down with a hoe. Lightly cover the entire planted area with a thin mulch of sand or compost. Sprinkle the area with a fine mist of water to avoid disturbing the seed or letting it wash away. The bed should be kept moist but not drenched to a muddy condition. Frequent light sprayings are in order until the first seedlings emerge.

E. Protecting the emerging seedlings

Once the plants begin to poke up out of the ground, don't assume that your work is over. Birds and other predators will finish off most of the delicious sprouts unless you've given them protection from the start. Bird netting, spunfiber cloth, or chicken wire—all available at most nurseries and hardware stores—may be necessary to keep your seedlings protected. Chicken wire or netting must be raised to a height of several inches over a bed or else birds will simply poke their beaks through to eat seeds and seedlings.

Lay these materials over the bed until the seedlings are as tall as your fist, with sturdy stems. By that time, you should be able to tell your seedlings from weeds. This is the time to reduce competition for your wildflowers. Otherwise, they will grow up stunted or malformed. Hand-pull the obvious alien weeds. In some cases, you may even want to thin out a cluster of natives, transplanting your removals to a more barren area. Continue occasional light watering until you see no more new seedlings sprouting.

F. Nursery stock transplanting

Many beautifully flowering trees, cacti, and shrubs are available from members of the Arizona Nursery Association and at botanical garden and arboretum plant sales. The native nursery stock found for sale is often that of one- to three-year-old plantings, providing you with a head start over what you could get by planting seeds. When buying these plants, select ones with healthy foliage, and if visible, root systems that do not appear too pot-bound or knotted. A vigorous looking plant can be easily transported and stored until you have prepared a hole for its transplanting.

For some woody perennials, you can root your own cuttings for later transplanting. Choosing the proper age of branches to cut, applying root hormones such as IBA (indolyl-butyric acid), and preparing a suitable rooting medium are all skills that can be learned by laymen. Although we encourage you to experiment with these vegetative propagation techniques, we realize that most home gardeners will seldom take on such a task themselves. Nevertheless, the Planting Guidelines gives species-by-species information for propagating many desert plants, in case you want to try. (See pages 104-107)

Choose a site where the plant can mature without growing up into the canopy of nearby trees and where its roots will not too fiercely compete with others nearby. If it is a sun-lover, make sure that it will not find itself in the shade of a faster-growing plant after a few years. If you do not want it to impinge on other plants, imagine how much shade it will cast when it reaches maturity. Then dig a hole roughly twice as wide as the root ball of the plant and a little more than twice as deep.

If you are aware that the tree is of a species that can be particularly deep-rooted, use a digging bar to loosen the soil or caliche to even a greater depth. For caliche-laden holes, some propagation specialists recommend ammonium sulfate application to the bottom of the hole to neutralize the calcium carbonate. However, caliche-dissolving chemicals are seldom as effective as a mechanical auger or good old-fashioned brute force in cracking hardpan.

Most nursery stock is sold in plastic containers that can be removed by turning the plant and pot upside-down, tapping the pot gently, and sliding the plant out. When cutting the nursery plant out of a metal can, be careful to keep as much of the soil within the root ball as possible. If the plant happens to be potbound, then uncoil its rootlets, cutting through any knots or tangles so that the loosened root mass can spread out once growth resumes.

Prepare the hole with a mix of screened native soil, peat, and compost to a depth where the exposed root ball will have its top just below the opening of the hole. Fill the hole with water to check for irregular drainage due to rodent holes and to provide a moist environment in which to place the plant. A slow-release fertilizer can be added into the fill at this time. Set the plant in place, then fill the rest of the hole, occasionally tamping the soil lightly to remove any air pockets.

When the hole is filled, make an earthen lip around its perimeter, and gently, slowly water this topsoil. If the plant is small, make a two-foot-tall cylinder of chicken wire to surround it for protection from rabbits or other animals that can damage it while still young. Extra shading and frequent watering may be required until signs of new growth become evident.

PHOTOGRAPHS BY DON B. STEVENSON

G. Low maintenance, not <u>no</u> maintenance

Obviously, most natural desert vegetation persists on rainfall, without supplemental irrigation. However, many of these plants are initially established during unusually wet years or may even require several consecutive wet years. Other plants grow only where storm runoff collects and soil moisture perseveres for much longer periods than on the surrounding desert slopes. Planted densely, desert natives do require some supplemental water.

In fact, many desert herbs and shrubs have growth rates directly proportional to the amount of water supplied them. That is to say, they will use water if it is provided to them but, if pressed, can survive on less than the amount that many introduced ornamentals require as a minimum.

The key factor for making the drought adaptations of native desert plants work for you is to water these natives on need, irrigating the area slowly and deeply. This encourages their roots to grow deeper and tap into a larger reservoir

DON B. STEVENSON

of soil moisture not touched by superficial evaporation from the surface.

Once deep roots are established, your plants will need to be watered less frequently and can endure on their own during extended periods when you are on vacation or otherwise away from home. Some rather large trees such as mesquite, paloverde, tree prickly-pear, and yucca can soon survive on their own with little or no supplemental water at all.

Watering is not the only maintenance required. Legumes tend to leave a litter of leaves and pods beneath their skirts. Depending upon the formality of your setting, you may want to rake or sweep these up during the season in which they appear. Paloverde seeds tend to sprout immediately after summer rains and may leave foot-tall seedlings within a month or two if its pods have not been swept away beforehand. Paloverdes and other trees also tend to shed leaves and small branches.

Unfortunately, not many native plants and trees are completely insect-and disease-resistant. Harvester ants are a constant summer problem. Beetles may infest branches, weakening them to the point of collapse during windstorms. Texas root rot and other diseases may "suddenly" kill handsome, mature shrubs. Particularly under typical conditions of cultivation and irrigation, desert plants may be more vulnerable to pestilence than they are in the wild. The University of

JAMES TALLON

Arizona College of Agriculture offers plant disease and plant pest "labs" several times a month in which researchers will identify your problem and suggest control methods or other solutions. The UofA's Cooperative Extension Service has produced tapes on "Plant Diseases" that are available through many public libraries in the state. Also, the Desert Botanical Garden's Plant Questions Hotline, (602) 941-1225, is available to those who live in the Phoenix area.

Overall, desert plants require no more maintenance than ones of other origin. It may be necessary to trim back the dried, dead stalks of desert ephemeral wildflowers after the season is over, but this holds true for other bedding plants as well. Pruning is necessary for straggly branches of mesquite as much as it is for those of mulberry or citrus. Dried-up annual wildflowers can be pulled up, shaken over the bed to release any seeds, and composted.

One difference with annual wildflowers, however, is that you can expect the wild desert species to self-propagate for many years following the initial release of a seed crop. And your original broadcasting of seeds may release sprouts for as many as three to five years. When they do appear and flower, take pleasure in them. Record the time of their arrival on a calendar. You can be sure that the duration and intensity of their flowering will be different every year.

DON B. STEVENSON

Take pleasure with your wildflowers while they are in bloom, for their fragrance and color change quickly.

Wildflower plantings for attracting birds and butterflies

Bird watching is one of the most important "non-destructive" outdoor recreation activities in Arizona.

Thousands of people come to southern Arizona canyons just to see species of hummingbirds, flycatchers, and trogons that reach into the United States from Mexico and at no other place. Yet the pleasures of watching birds, even some of the rarer ones, can start right at your kitchen window.

More than sugar-filled hummingbird feeders, than nestboxes and birdhouses, the most important investment you can make in attracting birds is through planting a diversity of flowering shrubs. These perennials provide not only food but cover and nesting sites as well. Some of the same plants attract butterflies.

In the species-by-species descriptions of wildflowers early in this book, we noted which of these provide nectar to hummingbirds. It may be worth summarizing the previously-noted species again: agaves, paloverdes, desert willows, ocotillos, gilias, chuparosas, and beardtongues. The following additional native species also attract hummingbirds: desert-honeysuckle; bird of paradise; hummingbird trumpets; penstemons, and sages. The Arizona-Sonora Desert Museum now features a hummingbird plant demonstration garden to encourage such plantings for wildlife.

Many other trees and shrubs give additional species of birds' food, shelter, and nesting sites. Desert hackberries, paloverdes, desert-willows, mesquites, Mexican elderberries, wolfberries, and saguaros offer some of the best multifaceted bird-attracters in our area. Even chollas and yuccas harbor certain birds which fail to obtain nesting sites in many other places. Anyone who has had the pleasure of watching an oriole family's nest slung from a yucca will be grateful for such a site in his own yard.

Planting shrubs to attract butterflies may seem far-fetched, but a growing number of Arizonans are doing just that. Among the recommended native trees and shrubs are desert broom, seep willow, elderberry, butterfly bush, bird of paradise, verbena, gilias, milkweeds, fairydusters, and desert senna.

TOM DANIELSEN

Chapter 9
Planting for food as well as beauty

Many of the same desert plants that offer food for birds can also intrigue the human palate. Saguaro fruit and prickly pear fruits, pads, and jellies demand good prices in gourmet food stores. Mesquite pod flour is being investigated as a commercial foodstuff; native desert peoples have already used it for thousands of years. Immature devil's claw pods are pickled and sold. Wild desert chiles demand a dry weight price of more than $30 a pound as a culinary spice. Wolfberries are still harvested in 10-pound quantities for syrup-making by members of the Gila River Indian Community in Arizona.

While all the food uses of desert wildflowers are too many to mention here, this sampler of recipes using the flowers has been prepared by Ruth Greenhouse, Desert Botanical Garden education associate. A book by Wendy Hodgson, Desert Botanical Garden botanist, offers an illustrated inventory of additional edible plants in the Sonoran Desert. It is just one more way to bring wildflowers into our lives.

Without thinking much about it, we regularly consume flowers. Artichokes, broccoli, and cauliflower can be found on almost every market produce counter. Flowers have always been part of the human diet, although their use has decreased in this century. Advances in food technology have allowed us to be less dependent on seasonally available foods.

Desert dwellers in the past have studied plant flowering patterns in order to take advantage of their dietary potential. If not to be eaten as is, flowers were a welcome event because they signaled the eminent arrival of edible fruits.

Flowers and flower buds serve different culinary purposes. Some can be used as a substantial part of a meal while others do better as a colorful, distinctive garnish. The recipes that follow are modern adaptations to the ways in which desert flowers were used by Native Americans in the past. The flowers discussed have been carefully selected and are safe to eat. Not all wildflowers are edible; some, like sacred datura, are poisonous.

All of the wildflowers recommended for use are easy to grow in your own garden. Collecting flowers (or other plant parts) from property other than your own without permission is illegal. (See summary of Arizona Native Plant Law, Chapter 2.) This practice also detracts from other people's enjoyment, as well as possibly seriously limiting the reproductive potential of the plant populations. Moreover, by growing and tending your own plants, you can be more confident that they are free from environmental contaminants.

Edible desert flowers: Gathering & preparation

1. *Collect flowers immediately after opening, in the early morning if possible.*

2. *Wash thoroughly but gently in cool water.*

3. *Drain and store between layers of paper towels to absorb excess moisture.*

4. *Refrigerate until used.*

Drying

1. *Line a cookie sheet with paper towels or brown paper.*

2. *Spread out clean whole flowers or blossoms in a single layer.*

3. *Set in a hot, dry place out of direct sunlight or place in oven at 150 degrees or less with the door slightly open.*

4. *Store in a covered container when thoroughly dry. Flowers must be brittle when crushed.*

(LEFT) *Fresh yucca petals, crisp in texture and delicate in flavor, are tasty combined with tender greens in a tossed salad with a vinaigrette dressing.* DON B. STEVENSON

yucca flowers

soaptree yucca - *Yucca elata*
Joshua-tree - *Yucca brevifolia*
Utah yucca - *Yucca utahensis*
Our Lord's candle - *Yucca whipplei*

The large, pale green yucca flowers are delicious both raw and cooked. The flowers of the narrow-leaf yuccas (mentioned above) are considered more palatable than the broad-leaf yuccas like banana yucca. Look for the blossoms from late April until late May. Unless you are going to crystallize the whole blossom as a garnish, only the petals should be used as the flower centers are bitter and a little tough.

Fresh yucca petals, crisp in texture and delicate in flavor, are tasty combined with tender greens in a tossed salad with a vinaigrette dressing. Yucca petals steamed or sauteed until tender, about 10 minutes, taste somewhat like celery and make a good side dish.

Dried yucca petals can be ground into flour and added to soups or stews. The thoroughly dried petals are easily pulverized with a small mortar and pestle. This flower "flour" can be stored in a covered container for future use.

JERRY JACKA

crystallized yucca flowers

(to decorate cakes and other desserts)

Eggwhite, beaten until slightly frothy

Small, soft brush

Fresh yucca blossoms, cleaned and dried

Finely granulated sugar

Line a cookie sheet with wax paper. Brush flowers, inside and out, with beaten egg white. Dip flowers into sugar. Place on cookie sheet. Dry in oven at 150 degrees or less with door slightly open. Store in covered container.

chilled yucca flower soup

---

4 cups clean yucca petals
(approx. 24 blossoms = 1 cup petals)

---

2 cups chicken broth

---

1 clove garlic

---

1 cup plain yogurt

---

1 cup sour cream

Dill weed

Boil yucca in chicken broth for 10 minutes. Cool. Puree in blender with garlic. Add yogurt and sour cream. Whirl just until blended. Chill.

Garnish with sprinkle of dill weed and fresh yucca petals.

chuparosa flowers

chuparosa - *Justicia californica*

The red, tubular flowers of chuparosa add a festive touch to recipes at a time of the year when other native flowers are generally not available. Chuparosa, dormant during hot seasons, blooms during the cool weather of fall and winter.

Raw chuparosa flowers are cool and crisp with a cucumber-like flavor. They make an attractive, tasty addition to fresh salads or a garnish for meat, fish, or poultry. Soups, fruits, or vegetables can be topped with a dollop of sour cream and a fresh chuparosa flower.

Chuparosa flowers are also good cooked and can be added to vegetable soups or stews.

Try freezing chuparosa flowers in individual ice cubes for a festive look.

CHARLES BUSBY

Arizona citrus-chuparosa salad

Butter or bib lettuce

Grapefruit and orange sections

Clean chuparosa flowers

Saguaro seed or poppy seed dressing

Arrange citrus sections on lettuce leaf. Top with saguaro seed dressing (recipe follows). Sprinkle with chuparosa flowers.

P. K. WEIS

saguaro seed dressing

½ cup mayonnaise
---
2 tablespoons honey
---
1 tablespoon lemon juice
---
1 tablespoon saguaro seeds
or poppy seeds

Mix them all together, and toss with your Arizona citrus-chuparosa salad.

barrel cactus flowers

barrel cactus - *Ferocactus wislizenii*

The flower buds and colorful yellow and red flowers of barrel cacti can be collected sometimes as early as January for use in recipes. Flowers and buds of the compas barrel, *F. acanthodes*, are generally avoided because they are bitter.

barrel blossom soup

3 tablespoons butter
3 tablespoons flour
2 cups chicken broth
2 cups chopped barrel blossoms
½ cup chopped onion
1 cup light cream or milk

Melt butter over medium heat. Stir in flour until smooth. Gradually add chicken broth, stirring constantly. Add remaining ingredients except cream. Simmer 10 minutes, stirring constantly. Add cream or milk. Reheat but do not boil.

buttered barrel buds

Drop barrel buds in boiling water to cover. Boil until tender, approximately 10 minutes. Drain and serve with butter, freshly ground black pepper, and a wedge of lime.

jellied barrel flowers

After pouring homemade jelly into sterilized jars, push a fresh, clean barrel cactus flower into the center of each jar. The hot jelly sterilizes and preserves the flowers. This is especially good with mesquite or cactus fruit jelly.

paloverde flowers

foothills paloverde - *Cercidium microphyllum*
blue paloverde - *Cercidium floridum*

In May, paloverdes decorate the desert with their dense yellow flowers. The foothills paloverde blossoms are creamy yellow while the blue paloverde blossoms are bright yellow. Delicate and mildly sweet, paloverde flowers should be gathered early in the morning as soon as possible after opening. They can be used fresh, either cooked, or as a garnish on salads or soups.

paloverde flower pudding

3 cups clean paloverde flowers
½ cup water
½ cup sugar
2 cups milk
2 tablespoons cornstarch
2 lightly beaten egg yolks or 1 well-beaten egg
2 tablespoons butter or margarine
¼ teaspoon salt

Simmer paloverde flowers in water until tender, approximately 10 minutes. Drain well and measure liquid. Add milk to make 1 cup. Puree blossoms in blender. In saucepan, blend sugar, cornstarch, and salt. Add milk and blossoms and cook while stirring over medium heat until thick and bubbly. Cook 2 minutes more, and remove from heat. Stir small amount of hot mixture into beaten egg. Return to hot mixture and cook 2 minutes more. Remove from heat, and add butter. Chill in dessert cups. Garnish with fresh paloverde blossoms.

paloverde pancakes

Add 1 cup of fresh paloverde flowers to your favorite pancake mix.

ocotillo flowers

ocotillo - *Fouquieria splendens*

The bright, scarlet tubular flowers of the ocotillo occur in dense clusters at the tips of the long stems in spring. Ocotillo buds as well as the sweet nectar-containing blossoms and the seeds are edible. Use a small rake or hooked stick to pull down the flexible branches so you can reach the blossoms. The flower clusters break off easily, but beware of the tough thorns on the branches.

JERRY JACKA

cucumber-ocotillo salad

2 cucumbers
2 tablespoons sugar
¼ cup white wine vinegar
¼ cup minced ocotillo buds
2 tablespoons minced parsley
Salt and pepper to taste

Slice cucumbers into thin slices. Add minced ocotillo buds and parsley. Add sugar and vinegar and toss. Chill 1 hour. Season with salt and pepper to taste.

DAVID BURCKHALTER

ocotillo flower punch

1 quart ocotillo blossoms
1 quart very warm (not hot) water
2 quarts 7-up
1 pint vanilla ice cream

Soak ocotillo blossoms in warm water overnight. Strain and chill. Mix ocotillo "tea" with chilled 7-Up. Add scoops of ice cream. Sprinkle with fresh ocotillo blossoms.

WILLARD CLAY

From an aesthetic viewpoint, the warm light and long shadows that occur at dawn and dusk provide dramatic as well as colorful photos in Organ Pipe Cactus National Monument. DAVID MUENCH

Planting guidelines

SPECIES	PAGE	FLOWER COLOR	USUAL MATURE PLANT HEIGHTS	COMMON METHODS OF PLANTING	NECESSARY SEED TREATMENTS	TIME OF PLANTING
sand-verbena *Abronia villosa*	50	Pink-purple to rose.	.5-1 ft.	Direct sowing; transplanting difficult.	No treatment needed.	Autumn
century plant *Agave parryi*	80	Red to pink bud, yellow when open.	2-3 ft., stalk 12-18 ft.	Seed in flats for transplanting.		Anytime
saiya *Amoreuxia palmatifida*	80	Yellow, red spot.	.3-1.2 ft.			Summer
desert-marigold *Bailey multiradiata*	50	Lemon yellow.	1-2 ft.	Seed in flats for transplanting; direct sowing.	No special treatment.	Anytime
fairyduster *Calliandra eriophylla*	50	Pink to rose.	2-4 ft.	Seed in pots for transplanting; direct sowing.	Soak in hot water.	Anytime
mariposa-lily *Calochortus kennedyi*	60	Orange-red to vermillion.	1-2 ft.	Seed in flats for transplanting.	No special treatment; use fresh seed.	Autumn
evening-primrose *Camissonia brevipes*	51	Yellow.	1-2 ft.	Direct sowing.	No special treatment.	Autumn
saguaro *Carnegiea gigantea*	70	Cream white.	Up to 30 ft.	Seed in flats for transplanting.	Light-dark alteration; warm temperature.	Spring-summer
blue paloverde *Cercidium floridum*	70	Yellow.	Up to 25 ft.	Seed in pots for transplanting; semi-hardwood treated with indolyl-butyric acid (IBA) rooting hormone and misted.	Eliminate weevil-infested seed.	Spring-summer
foothills paloverde *Cercidium microphyllum*	70	Yellow.	Up to 20 ft.	Seed in pots for transplanting; semi-hardwood treated with IBA rooting hormone and misted.	Eliminate weevil-infested seed.	Spring-summer
pincushion plant *Chaenactis stevioides*	51	Cream to lemon-yellow.	.5-1 ft.	Direct sowing.	Not known.	Autumn
desert-willow *Chilopsis linearis*	80	White, lavender or purple with yellow spots.	10-30 ft.	Seed in pots for transplanting; semi-hardwood cuttings treated with IBA rooting hormone and misted; dormant winter cuttings.	Soak fresh seed in water for 2-3 hours prior to sowing.	Spring
coyote gourd *Cucurbita digitata*	81	Yellow.	Vines 4-12 ft. long.	Direct sowing; transplanting of dormant roots.	Allow to ripen in fruit for 45-60 days following flowering.	Spring-summer
sacred datura *Datura meteloides*	71	White with purple or lavender throat.	2-3 ft.	Seed in flats for transplanting; direct sowing.	No special treatment.	Spring-summer
larkspur *Delphinium scaposum*	60	Blue-bronze or purple, and white.	1-2 ft.	Seed in pots for transplanting.	No special treatment for fresh seed.	Early spring
strawberry hedgehog *Echinocereus engelmannii*	60	Reddish-purple to magenta.	.5-1.5 ft.	Seed in pots; cuttings.	Sow onto soil surface after treating with fungicide. Keep in light at 70°-75° F. or higher, in moist soil, and cover with plastic bag.	Spring-summer
brittlebush *Encelia farinosa*	51	Yellow.	1-3 ft.	Seed in pots for transplanting; direct sowing.	No special treatment.	Spring-summer
coral bean *Erythrina flabelliformis*	81	Red.	2-5 ft.	Seed in pots for transplanting; semi-hardwood cuttings taken in summer or fall, misted.	Scarify seedcoat with knife, or soak in concentrated sulfuric acid 20-45 minutes.	Spring
poppy *Eschscholtzia mexicana*	61	Gold to orange; rarely white or pink.	.5-2 ft.	Direct sowing; transplanting difficult.	No special treatment.	Autumn-winter

Planting guidelines

SPECIES	PAGE	FLOWER COLOR	USUAL MATURE PLANT HEIGHTS	COMMON METHODS OF PLANTING	NECESSARY SEED TREATMENTS	TIME OF PLANTING
barrel cactus *Ferocactus wislizenii*	81	Tangerine-orange.	2-4 ft., rarely 10 ft.	Seed in pots for transplanting.	Sow onto soil surface after treating with fungicide. Keep in light at 70°-75° F. or higher, in moist soil, and cover with plastic bag.	Spring-summer
ocotillo *Fouquieria splendens*	61	Maroon-red.	7-10 ft.	Seed in flats or pots for transplanting; place stem cuttings in well-drained soil. Spray stem cuttings once daily.	Sow onto soil surface after treating with fungicide. Do not overwater.	Summer
blanketflower *Gaillardia arizonica*	52	Yellow.	.5-1.5 ft.	Seed in flats for transplanting; place stem cuttings in well-drained soil.	No special treatment.	Autumn-winter
desert-sunflower *Geraea canescens*	61	Yellow.	1-2 ft.	Direct sowing.	No special treatment.	Autumn
gilia *Gilia latifolia*	62	Pink to lavender with yellow throat.	.5-1 ft.	Direct sowing.	No special treatment.	Autumn
Goodding's-verbena *Glandularia gooddingii*	52	Mauve to lavender.	.5-2 ft.	Direct sowing.	No special treatment.	Winter-spring
wild cotton *Gossypium thurberi*	82	Whitish to pale pink.	4-8 ft.	Seed in pots for transplanting; direct sowing.	Scarify seedcoat with knife or soak in hot water.	Spring-summer
snowy sunflower *Helianthus niveus*	82	Yellow.	1-5 ft.	Direct sowing.	Scarify seed. Remove "shell" or achene before sowing true seed.	Autumn-spring
ajo-lily *Hesperocallis undulata*	52	White.	.5-1 ft.	Direct sowing; bulbs for direct planting.	No special treatment for fresh seed.	Autumn
desert rosemallow *Hibiscus coulteri*	82	White to yellow.	3-4 ft.	Seed in flats for transplanting; softwood cuttings treated with IBA rooting hormone and misted.	Soak seeds in warm water for 24 hours.	Spring
chuparosa *Justicia californica*	53	Red.	2-5 ft.	Seed in pots for transplanting; cuttings.	No special treatment.	Spring
Arizona caltrop *Kallstroemia grandiflora*	83	Orange with red veins.	1-2.5 ft.	Direct sowing.	No special treatment.	Spring-summer
creosotebush *Larrea tridentata*	53	Yellow.	4-7 ft.	Direct sowing.	Hull seeds or sand off one end; soak 4-9 hours in deionized water.	Anytime
goldfields *Lasthenia chrysostoma*	62	Yellow.	Up to 1 ft.	Direct sowing.	No special treatment.	Autumn
tidytips *Layia glandulosa*	62	White or pink ray-like petals; yellow or orange disks.	Up to 1 ft.	Direct sowing.	No treatment needed.	Autumn-winter
bladderpod mustard *Lesquerella gordoni*	53	Yellow.	Up to 1 ft.	Direct sowing.	No special treatment.	Winter-spring
lupine *Lupinus sparsiflorus*	54	Violet-blue.	.5-2 ft.	Direct sowing.	Scarify seed and/or soak in hot water 4-8 hours.	Autumn
pincushion cactus *Mammillaria microcarpa*	83	Pink to lavender.	Up to .5 ft.	Seed in flats for transplanting; vegetative cuttings.	Dust fungicide onto seed, sprinkle onto fine gravel atop cactus mix soil, keep moist or in mist.	Spring-summer

Planting guidelines

SPECIES	PAGE	FLOWER COLOR	USUAL MATURE PLANT HEIGHTS	COMMON METHODS OF PLANTING	NECESSARY SEED TREATMENTS	TIME OF PLANTING
blackfoot daisy *Melampodium leucanthum*	63	White with purple veins.	.5-2.5 ft.	Direct sowing.	No special treatment.	Autumn
blazingstar *Mentzelia involucrata*	63	Cream or lemon.	1.5-2.5 ft.	Direct sowing.	No special treatment.	Autumn
ironwood *Olneya tesota*	71	Purple and white.	8-25 ft.	Seed in pots for transplanting.	Scarify seed and soak in hot water for 4-8 hours.	Spring-summer
buckhorn cholla *Opuntia acanthocarpa*	71	Red, orange or greenish-yellow.	4-7 ft.	Seed in flats for transplanting; stem joints for vegetative transplants.	Scarify seed, treat with fungicide, and plant atop cactus mix soil; cover lightly with gravel and keep moist.	Spring-summmer
pancake-pear *Opuntia chlorotica*	72	Yellow.	3-6 ft.	Seed in flats for transplanting; stem pads for vegetative transplants.	Scarify seed, treat with fungicide, and plant atop cactus mix soil; cover lightly with gravel and keep moist.	Spring-summmer
Engelmann's prickly-pear *Opuntia engelmannii*	72	Yellow to pink.	3-6 ft.	Seed in flats for transplanting; stem joints for vegetative transplants.	Scarify seed, treat with fungicide, and plant atop cactus mix soil; cover lightly with gravel and keep moist.	Spring-summmer
jumping cholla *Opuntia fulgida*	83	Pink.	4-8 ft.	Seed in flats for transplanting; stem joints for vegetative transplants.	Scarify seed, treat with fungicide, and plant atop cactus mix soil; cover lightly with gravel and keep moist.	Spring-summmer
purple prickly-pear *Opuntia violacea*	72	Yellow.	2-4 ft.	Seed in flats for transplanting; stem pads for vegetative transplants.	Scarify seed, treat with fungicide, and plant atop cactus mix soil; cover lightly with gravel and keep moist.	Spring-summmer
owl-clover *Orthocarpus purpurascens*	63	Purplish-pink to red, with yellow spots.	.5-1 ft.	Direct sowing.	No special treatment.	Autumn
night-blooming cereus *Peniocereus greggii*	84	White with red and green-tinged mid-ribs.	Up to 7 ft., often less than 3 ft., rarely 11 ft.		Light-dark alteration; warm temperatures.	Spring-summer
beardtongue *Penstemon parryi*	64	Rose-magenta.	1-3 ft.	Seed in pots for transplanting; direct sowing.	No special treatment; prefers light.	Autumn
scorpionweed *Phacelia crenulata*	54	Violet-purple, rarely white.	1-3 ft.	Direct sowing.	Scarify seed, soak 4-6 hours in hot water; inhibited by light if seed-coat is intact.	Autumn
phlox *Phlox tenuifolia*	64	White and lavender.	1.5-3 ft.	Direct sowing.	No special treatment. Needs partial shade or reduced light.	Autumn
devil's claw *Proboscidea parviflora*	84	Reddish-purple to white, with pink and yellow in throat.	1.5-5 ft.	Direct sowing.	Requires light-dark alteration; scarify seed-coat before sowing.	Spring-summer
mesquite *Prosopis velutina*	73	Yellow.	8-30 ft.	Seed in pots or sleeves for transplanting; cuttings treated with IBA rooting hormone and misted.	Fumigate seed upon collection to kill weevils; scarify seedcoat then soak in concentrated sulfuric acid for 1/2 hour or less.	Spring-summer

Planting guidelines

SPECIES	PAGE	FLOWER COLOR	USUAL MATURE PLANT HEIGHTS	COMMON METHODS OF PLANTING	NECESSARY SEED TREATMENTS	TIME OF PLANTING
desert-chicory *Rafinesquia neomexicana*	64	White.	.5-2 ft.	Direct sowing.	No special treatment.	Autumn
chia *Salvia columbariae*	65	Blue to indigo.	.5-2 ft.	Direct sowing.	No special treatment.	Autumn-winter
elderberry *Sambucus mexicana*	73	White.	10-30 ft.	Seed in pots for transplanting; softwood or semi-hardwood cuttings treated with IBA rooting hormone and misted.	Use fresh seed. Clean pulp off seed immediately and air dry in sun. Scarify seed in sulfuric acid 10-15 min. or warm-stratify for 2 months at temperatures of 70°-80° F.	Spring-summer
desert senna *Senna covesii*	84	Rust-yellow.	1-2 ft.	Check seed source	No special treatment.	Spring
jojoba *Simmondsia chinensis*	54	Yellow-green.	4-7 ft.	Seed in pots or sleeves for transplanting; stem cuttings treated with IBA rooting hormone, misted and kept over low heat.	No special treatment.	Summer
globemallow *Sphaeralcea ambigua*	55	Peach, apricot or grenadine.	2-6 ft.	Direct sowing.	No special treatment.	Autumn
trumpetflower *Tecoma stans*	73	Yellow.	3-9 ft.	Seed in flats for transplanting; direct sowing; fall semi-hardwood cuttings treated with IBA rooting hormone and misted.	No special treatment.	Spring-summer
crownbeard *Verbesina encelioides*	65	Yellow and orange.	1-3 ft.	Direct sowing.	No special treatment.	Autumn
Joshua-tree *Yucca brevifolia*	65	Greenish-white.	10-25 ft.	Seed in flats for transplanting.	Treat seed with fungicide and place atop cactus mix soil, then cover lightly with coarse gravel, and keep moist.	Spring-summer
soaptree yucca *Yucca elata*	74	Creamy white.	5-15 ft.	Seed in flats for transplanting; rhizome cuttings can be transplanted.	Treat seed with fungicide and place atop cactus mix soil, then cover lightly with coarse gravel, and keep moist.	Spring-summer
desert zinnia *Zinnia grandiflora*	85	Yellow.	.5-.75 ft.	Check seed source.	Not known.	Autumn

---

AUTHORS' NOTE

Additional information for planting may be obtained from native seed and plant sources in Arizona and neighboring states. This book features wildflowers for which seeds and plants are commercially available. A listing of these suppliers may be obtained from the Desert Botanical Garden, 1201 North Galvin Parkway, Phoenix, Arizona 85008. Visitors there may obtain from the library a copy of the "seed/plant sources list" that is updated yearly.

References

Baumgardt, John Philip *How to Identify Flowering Plant Families: A Practical Guide for Horticulturists and Plant Lovers.* Portland: Timber Press, 1982. 269 p.

Dillard, Annie *Teaching a Stone To Talk.* New York: Harper and Row, 1983. 192 p.

Dodge, Natt N. *Flowers of the Southwest Deserts.* Tucson: Southwest Parks and Monuments Association, 1985. 136 p.

Duffield, Mary Rose and Warren Jones *Plants for Dry Climates: How to Select, Grow and Enjoy.* Tucson: H.P. Books, Fisher Publishing, 1981. 176 p.

Hodgson, Wendy Caye *Edible Native and Naturalized Plants of the Sonoran Desert North of Mexico.* Ann Arbor: University Microfilms International, 1985. 502 p.

Kearney, Thomas H. and Robert H. Peebles *Arizona Flora.* Berkeley: University of California Press, 1960. 1085 p.

Lehr, J. Harry *A Catalogue of the Flora of Arizona.* Phoenix: Desert Botanical Garden, 1982. 203 p. and supplements.

McGinnies, William G. *Flowering Periods for Common Desert Plants: Southwestern Arizona.* Tucson: University of Arizona Press, 1986. Poster.

Nabhan, Gary Paul *The Desert Smells Like Rain: A Naturalist in Papago Indian Country.* San Francisco: North Point Press, 1982. 148p.

Gathering the Desert. Tucson: University of Arizona Press, 1985. 209 p.

Saguaro. Tucson: Southwest Parks and Monuments Association, 1986. 73 p.

Niehaus, Theodore F. *A Field Guide to Southwestern and Texas Wildflowers.* Boston: Houghton Mifflin, 1984. 449 p.

Nokes, Jill *How to Grow Native Plants of Texas and the Southwest.* Austin: Texas Monthly Press, 1986. 404 p.

Shreve, Forrest and Ira L. Wiggins *Vegetation and Flora of the Sonoran Desert.* Stanford: Stanford University Press, 1964. 1740 p.

Spellenberg, Richard *The Audubon Society Field Guide to North American Wildflowers: Western Region.* New York: Alfred A. Knopf, 1979. 862 p.

Wilson, William H.W. *Landscaping With Wildflowers and Native Plants.* San Francisco: Chevron Chemical Company, Ortho Books, 1984. 93 p.

Young, James A. and Cheryl G. Young *Collecting, Processing, and Germinating Seeds of Wildland Plants.* Portland: Timber Press, 1986. 236 p.

Index and glossary

Index and glossary

Index and glossary

Index and glossary